QUEST *for the* MOON

AND OTHER STORIES

THREE DECADES OF ASTRONAUTS IN SPACE

ANNE WILKES TUCKER WITH DENNIS IVY

THE MUSEUM OF FINE ARTS, HOUSTON

Quest for the Moon and Other Stories: Three Decades of Astronauts in Space was published in conjunction with the exhibition organized by The Museum of Fine Arts, Houston, and held December 18, 1994, through February 5, 1995.

Support for this exhibition has been provided by Target Stores, The Blount Foundation, Rockwell International, Houston Endowment Inc., The Color Place, and Dr. and Mrs. Joseph P. Allen.

Publications Director: Diane Planer Lovejoy
Graphic Design: Julie Graf
Printing: The Beasley Co., Inc.
Color separations: Color Separations, Inc.

ISBN 0-89090-065-5

Printed in the United States of America

Cover: Mission: Space Shuttle 62, March 4–18, 1994. *Launch of Space Shuttle Columbia at 8:53 a.m. (E.S.T.), March 4, 1994.* Dye-transfer photograph. The Museum of Fine Arts, Houston; The Target Collection of American Photography; museum purchase with funds provided by Target Stores.

Lenders to the Exhibition

Frish Brandt

Fondren Library, Rice University, Houston

Joan Fontcuberta

Fraenkel Gallery, San Francisco

Gernsheim Collection, Harry Ransom Humanities Research Center, The University of Texas at Austin

Harvard College Observatory, Cambridge, Massachusetts

G. Ray Hawkins Gallery, Santa Monica

Frank P. Herrera

Robert Koch Gallery, San Francisco

Association des Amis de J. H. Lartigue, Paris

Harry Lunn, Ltd., New York

Robert Mann Gallery, New York

Richard Misrach

Ikko Narahara

Lyndon B. Johnson Space Center of the National Aeronautics and Space Administration, Houston

Galerie Pons, Paris

Richard Ross

Manuel Sendón

Robert Shlaer

Andrew Smith Gallery, Santa Fe

Jim Stone

Walker Arts Center, Minneapolis

West End Gallery, Houston

Daniel Wolf, Inc., New York

Zabriskie Gallery, New York

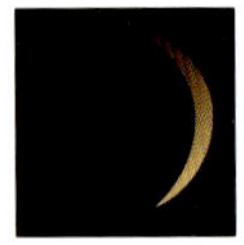

Artists Represented

Ansel Adams
Wallace Berman
Michael Bishop
Paul Caponigro
James Carpenter
Edward Sheriff Curtis
Warren de la Rue
Rafail Diament
Elliott Erwitt
Joan Fontcuberta
Henry Frères
Frank P. Herrera
Ton Huijbers
Samuel Dwight Humphrey
Mark Klett
Jacques-Henri Lartigue
Robyn McDaniels
Richard Misrach
Ikko Narahara
James Nasmyth
Negretti & Zambra Company
Kathleen Packlick
Luis González Palma
Puiseux and Le Morvan
Robert Rauschenberg
Richard Ross
Manuel Sendón
Robert Shlaer
Alfred Stieglitz
Joel Sternfeld
Jim Stone
U. S. Astronaut Corps
Thierry Urbain
John Adams Whipple
Garry Winogrand
Max Yavno

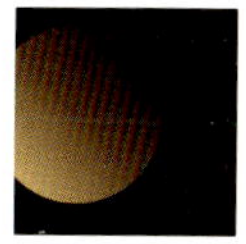

Table of Contents

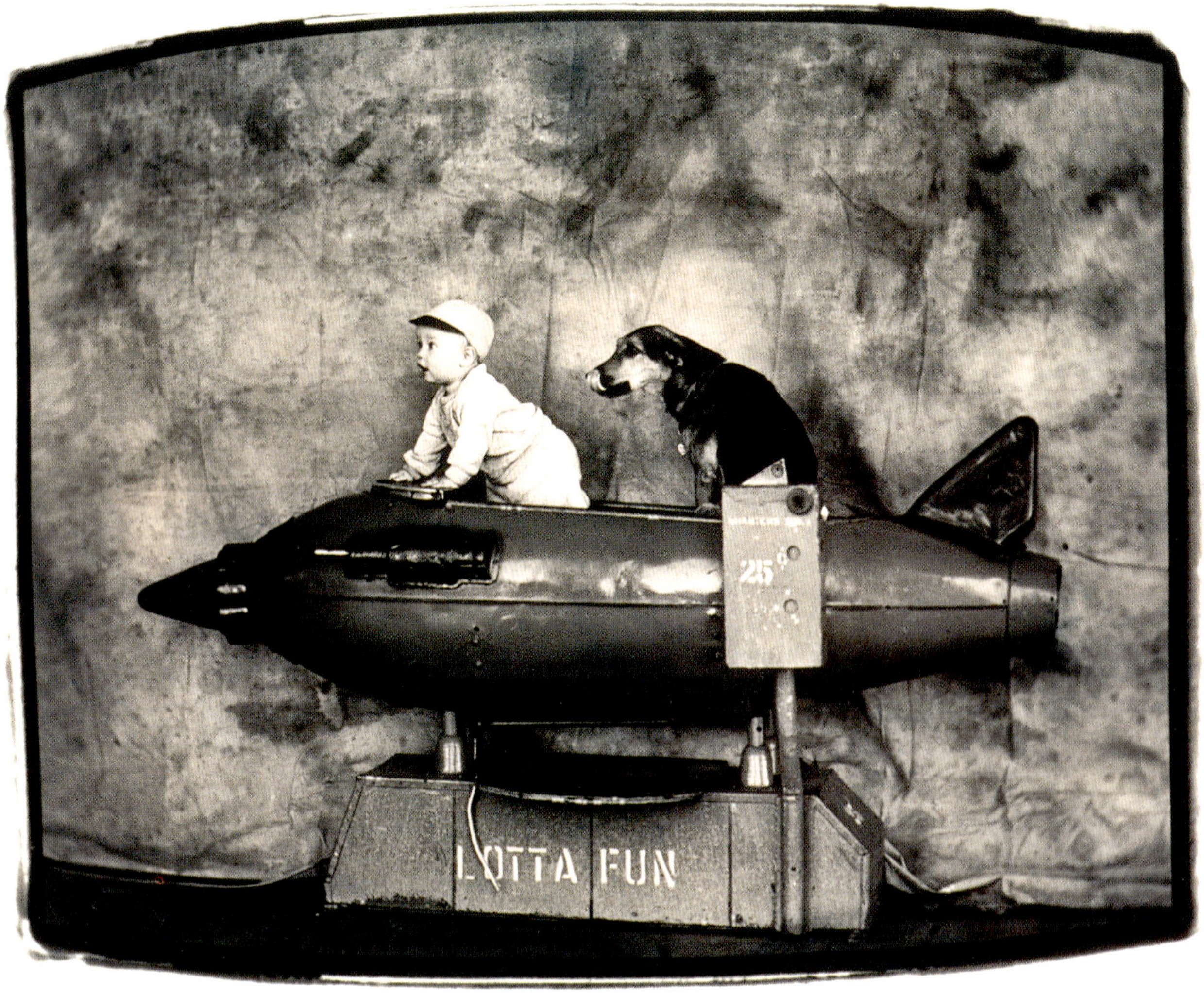

Robyn McDaniels. *Oscar Takes Larry to the Moon.* 1993.
Gelatin silver photograph. The Museum of Fine Arts, Houston; museum purchase with funds provided by Clinton T. Willour in honor of Frish Brandt.

Acknowledgments

Quest for the Moon and Other Stories: Three Decades of Astronauts in Space is the eighth in a series of exhibitions at the Museum of Fine Arts, Houston funded by Target Stores. The corporation generously provided funds for the museum to purchase many of the works on view as well as contributed to the costs of the exhibition and this catalogue. The exhibition is also supported with funds provided by The Blount Foundation, Rockwell International, Houston Endowment Inc., The Color Place, and Dr. and Mrs. Joseph P. Allen.

This exhibition benefited significantly from the curatorial assistance of Dr. Joseph P. Allen, a NASA astronaut, and Dennis Ivy, a Houston-based photographer and master printer. Ivy reviewed thousands of images from NASA's vast photographic collection both for the initial portfolio, *NASA: A Quarter Century of Space Exploration,* which became the core of the exhibition, and for photographs made by NASA in the last decade. He and other experts culled the first round of selections, and then Ivy produced the master prints using materials generously supplied by NASA. He published the portfolio in 1991.

The experts who worked with Ivy to select images for the portfolio included James Dean, a former director of the NASA art program and a curator of the Smithsonian Institution, Washington, D.C.; George Krause, a professor in the photography department at the University of Houston; Roy Flukinger, curator of the Gernsheim Collection of the Harry Ransom Humanities Research Center, the University of Texas at Austin; and Anne Wilkes Tucker, Gus and Lyndall Wortham Curator of Photography, the Museum of Fine Arts, Houston. The NASA photographs that date after 1984 were selected by Ivy, Tucker, and Allen.

I would also like to express the museum's gratitude to the National Aeronautics and Space Administration for graciously facilitating our utilization of its archival resources and for lending a lunar sample to the exhibition. Key individuals who deserve special thanks are Mike Gentry, Mary Louise Schmid, James Hartfield, and Louis Parker of NASA's Johnson Space Center; and Barry B. Schroder, Gary Seloff, and Mary N. Wilkerson of RMS Photographic Services at the Johnson Space Center. Thanks go also to Carolyn Windle of The Color Place, Don Avrut, Sallye Webster, Robert D. Stecker, and G. S. Henson, Jr.

Although the majority of works in the exhibition are in the museum's collection, there are other key images that numerous institutions and individuals have made available to the Museum of Fine Arts. These lenders are recognized on page 3, and we greatly appreciate their willingness to enhance this exhibition with their works.

Anne Wilkes Tucker would like particularly to thank Robert Mann; archivists Joseph Struble and Brenda Corbin; historians M. Susan Barger, Françoise Heilbrun, Martha Hazen, and John Wood; photographer Robert Shlaer; curatorial assistant R. Eric Davis, and museum trustee Pampa Trotti for their research assistance. Mark Petr, a former curatorial assistant in the photography department, also assisted with research and wrote two of the entries for this catalogue.

Peter C. Marzio
Director
The Museum of Fine Arts, Houston

Introduction

Throughout history, the moon has meant many things to many people. Initially, it was humankind's only ally against the dangers of the night, a means of marking the passage of time, a god, an inspiration to artists, and an icon of love. In the ancient world, the connection between the measurement of time as observed in lunar and seasonal cycles and early myth and religion was almost universal. As people became aware of the need to measure time more accurately, they began to observe celestial phenomena with greater care. Their early observations were the precursors of the new science of astronomy.

The first important observations of the moon with telescopes were made in 1609 by the Italian Galileo Galilei (1564-1642) and the Englishman Thomas Harriot (1560-1621). Galileo observed the moon through a telescope of his own construction and then produced and published five primitive drawings of the lunar surface detailing craters, "seas," and mountains. Thereafter, the science of optics and construction of telescopes improved rapidly; lunar observations and records progressed steadily as well.

With the invention of photography in 1839, lunar astronomers could document what they observed with increasing effectiveness. Louis Jacques Mandé Daguerre (1789-1851) of Paris, one of photography's inventors, attempted the first photograph of the moon in 1839, but failed owing to the insensitivity of his materials. John William Draper (1811-1882), an American astronomer, made a series of lunar daguerreotypes in 1840. Within a decade, these were followed by others produced with improved telescopes, most notably by Professor William Cranch Bond (1789-1859) and photographer John Adams Whipple (1822-1891) at the Harvard College Observatory. The photographic atlas of the moon produced in Paris by Loewy and Puiseux, with the assistance of Le Morvan, between 1896 and 1910 remained the most comprehensive work on the subject until 1960.

Cameras launched by rockets achieved the next great advances in moon imagery. The first cameras propelled by rocket were aimed at the Earth. The U.S. Army launched a DeVry 35mm motion-picture camera in 1946 on a captured German V-2 rocket; it reached a height of sixty-five miles. The first detailed photographs of the moon's dark side were taken by USSR Lunar 3 in 1959. In July 1964, the USA Ranger VII impacted on the moon and sent back more than 4,000 high-quality photographs to Earth. Probes that orbited the moon and sent back high-resolution images soon followed. Photographs made by men on the moon represent the highest achievements of lunar documentation.

Knowledge has not diminished our fascination with the moon as a constant nightly presence and yet a mysterious force in our lives. Few people still worship the moon, but its eerie beauty and iconic qualities assure its continued presence in both the visual and verbal arts of many cultures. This exhibition celebrates the moon both as a goal for the greatest scientific exploration of our century and as an entity of historical and current humanistic importance.

The exhibition also focuses on photography's role in our desire to understand the moon; the photographs were chosen for their artistic as well as for their scientific merit. Since this is a modest exhibition that does not attempt to be comprehensive, many photographs chronicling major moments in lunar observation and in the history of the space program are not shown, such as photographs from John Glenn's pioneering orbit around the Earth in 1962. Our fundamental goal is to convey the qualities of beauty, wonder, and artistry—as the medium of photography is uniquely capable of expressing—that have stimulated our fascination with the moon and propelled our quest to reach it.

Anne Wilkes Tucker
Gus and Lyndall Wortham Curator of Photography
The Museum of Fine Arts, Houston

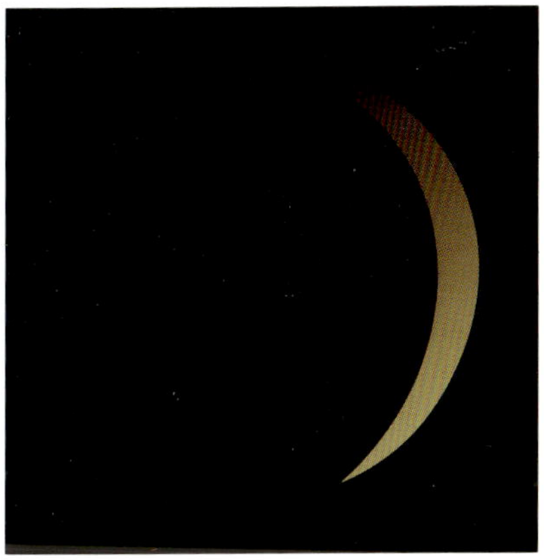

NINETEENTH-CENTURY PHOTOGRAPHS *of the* MOON

The Astronomical Daguerreotype

When two of photography's inventors, Nicéphore Niépce (1765-1833) and Louis Jacques Mandé Daguerre (1789-1851), sought support from the French government for their new invention, they were lucky to obtain the endorsement of François Dominique Arago (1786-1853). In his position as the director of the Paris Observatory, the secretary of the Académie des Sciences, and a member of the French Parliament, Arago was ideally suited to promote these photographers' accomplishments. With his encouragement, the French government purchased the rights to Niépce's and Daguerre's invention, and made the process available to the world.

Upon Arago's request, Daguerre captured a faint image of the moon on a daguerreotype plate, but the process he employed was too insensitive and he lacked sufficiently sophisticated equipment to be successful. As historian M. Susan Barger has explained:

> Astronomical daguerreotypy required the mastery of a combination of both photographic and mechanical skills. To make an astronomical daguerreotype, the telescope is made into a camera by positioning the unexposed plate in a holder at the eye-piece of the telescope.... Further, to compensate for the movement of the Earth relative to the other heavenly bodies, the telescope had to have an accurate tracking mechanism or clock drive to move it during exposures; otherwise, images would appear smeared, blurred, or misshapen on the plate.[1]

There were further problems caused by the refraction of light by the telescope lenses.

The first successful lunar daguerreotypes were made in 1840 by the American astronomer and daguerreotype portraitist John William Draper (1811-1882). Within a decade, these early, successful examples were followed by others produced with improved telescopes, most notably by photographer John Adams Whipple (1822-1891) and Professor William Cranch Bond (1789-1859), director of the Harvard College Observatory, which then owned the largest telescope in the world, the Great Refractor. In 1848, only four months after the telescope's completion, Bond and Whipple began their attempts to photograph the sun and moon. Over the next decade, Whipple (and his partners) and Bond and his son, George Phillips Bond (1825-1865), made nearly seventy daguerreotypes. They sought to produce images that were beautiful as well as acute and accurate. Widely published and exhibited in Europe, their work inspired others to make astronomical photographs.

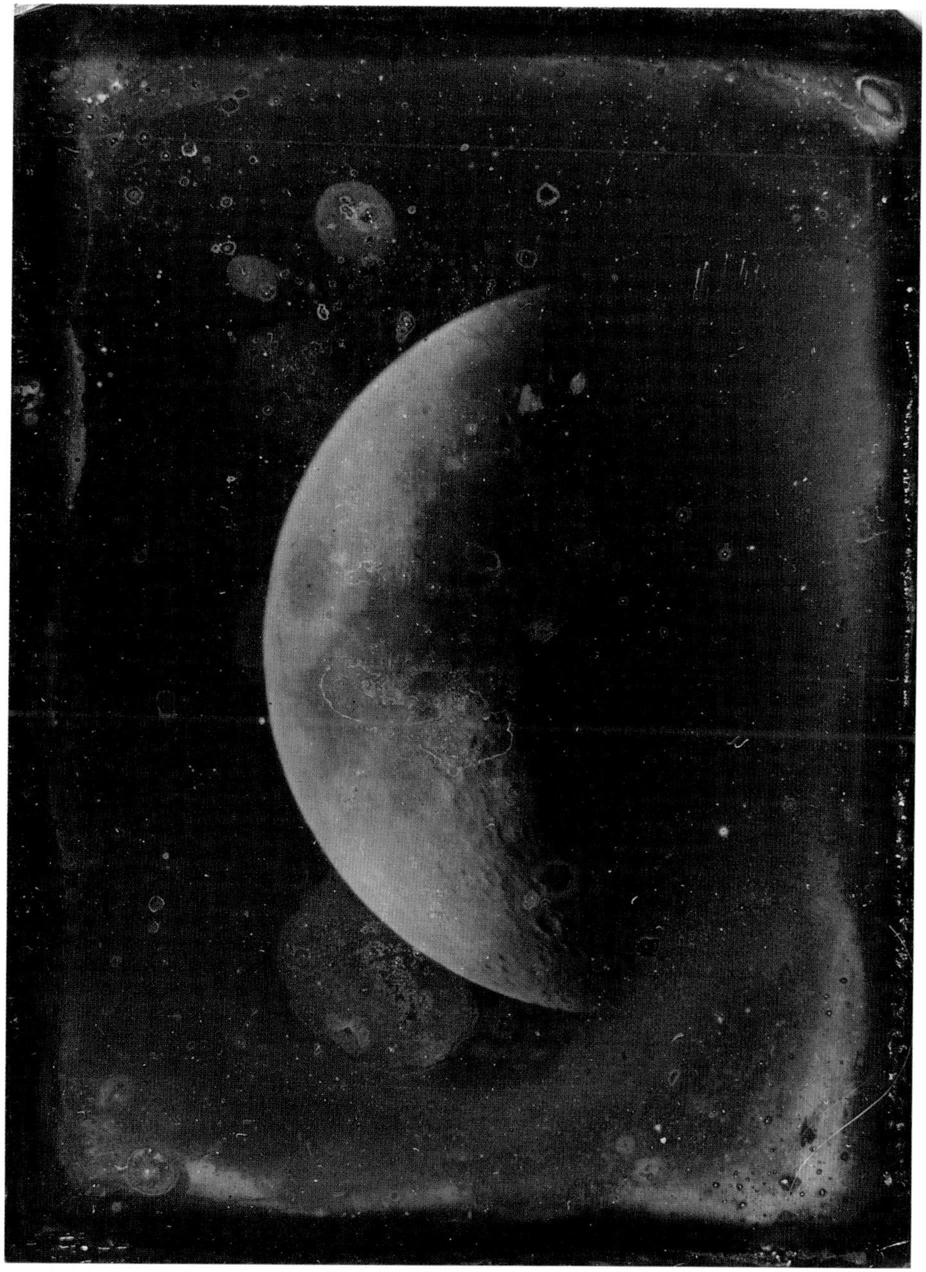

John Adams Whipple. *The Moon*. February 26, 1852. Daguerreotype.
Courtesy of Harvard College Observatory, Cambridge, Massachusetts.

Warren de la Rue

In 1851, inspired by the lunar daguerreotypes produced by Whipple and Bond, Warren de la Rue (1815-1889), an English astronomer and chemist, took up astronomical photography. Working at his private observatory in Cranford, England, he chose to use the wet collodion process for producing photographic negatives, rather than to make daguerreotypes. He lectured, published accounts of his photographic works, and organized teams of astronomers and photographers to photograph celestial events, such as eclipses. His efforts in both lunar and solar photography were unsurpassed for many years. De la Rue became known as the "Father of Astronomical Photography."

James Nasmyth

In 1874 a Scottish engineer, James Nasmyth (1808-1890), and James Carpenter, an astronomer at the Royal Observatory in Greenwich, published *The Moon, Considered as a Planet, a World, and a Satellite*. Illustrated with an unusually wide variety of high-quality reproduction processes, including wood engravings, lithographs, photogravures, and woodburytypes, the book culminated forty years of lunar observation. Over the next decade, the book was published in numerous editions in several languages, both in Europe and in the United States.

To illustrate his points, Nasmyth made photographs and highly detailed drawings and paintings of what he observed through his telescope. He also made plaster models of the moon that he photographed in raking light to faithfully reproduce the lunar effects of light and shadow. To promote his conviction that the moon's surface was formed by volcanos, Nasmyth even drew steam on a reproduction plate before printing it. This attempt to convey the sense of an active erupting volcano is thought to be the first example of "space simulation." Nasmyth also photographed the back of a hand and a shriveled apple to illustrate the origin of certain mountain ranges by shrinkage of the globe.

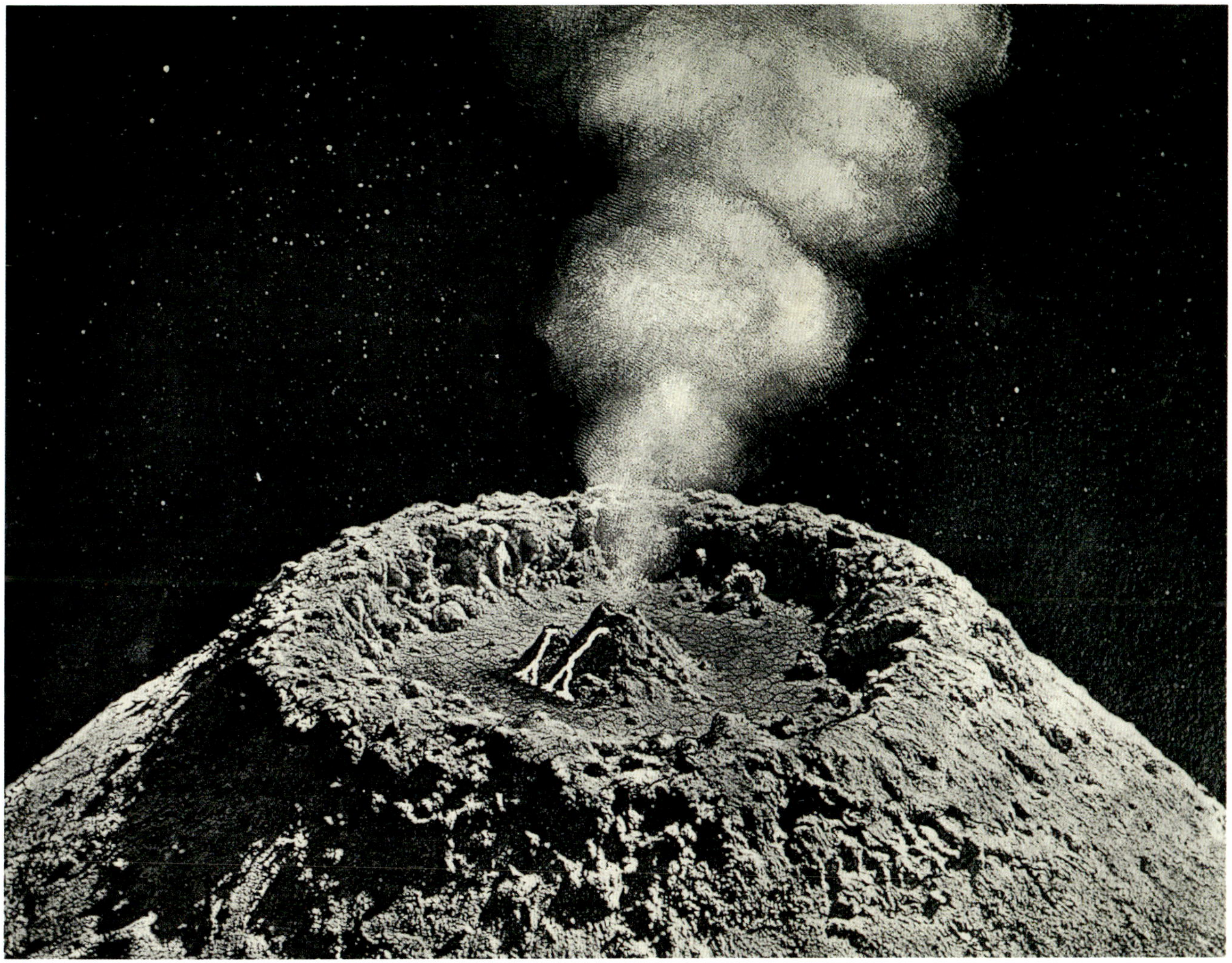

James Nasmyth. *Crater of Vesuvius*. 1864. Woodburytype photograph. The Museum of Fine Arts, Houston; gift of Clinton T. Willour in honor of Peter C. Marzio.

The Paris Observatory

From the moment of photography's invention, the staff of the Paris Observatory recognized the medium's unique ability to record and advance studies of the moon, the planets, and the stars. The first daguerreotype of the sun was made by Jean Bernard Léon Foucault (1819-1868) and Hippolyte Fizeau (1819-1896) at the observatory in April 1845. Other distinguished astronomical photographs were made by the astronomers Paul Pierre (1849-1901) and Prospère Henry (1848-1902) and, slightly later, by Loewy and Puiseux. The Henry brothers were not only skilled astronomers but also "optical geniuses" who designed a special camera for the observatory telescope. Between 1896 and 1910, Loewy and Puiseux published seventy-one photographic plates in a twelve-part atlas of the moon, which remained the most comprehensive work on the subject for fifty years.

Henry Frères (Paul Pierre Henry and Prospère Henry).
Lunar Photograph, Paris Observatory, March 20, 1890.
Albumen photograph. Courtesy of Harry Lunn, Ltd., New York.

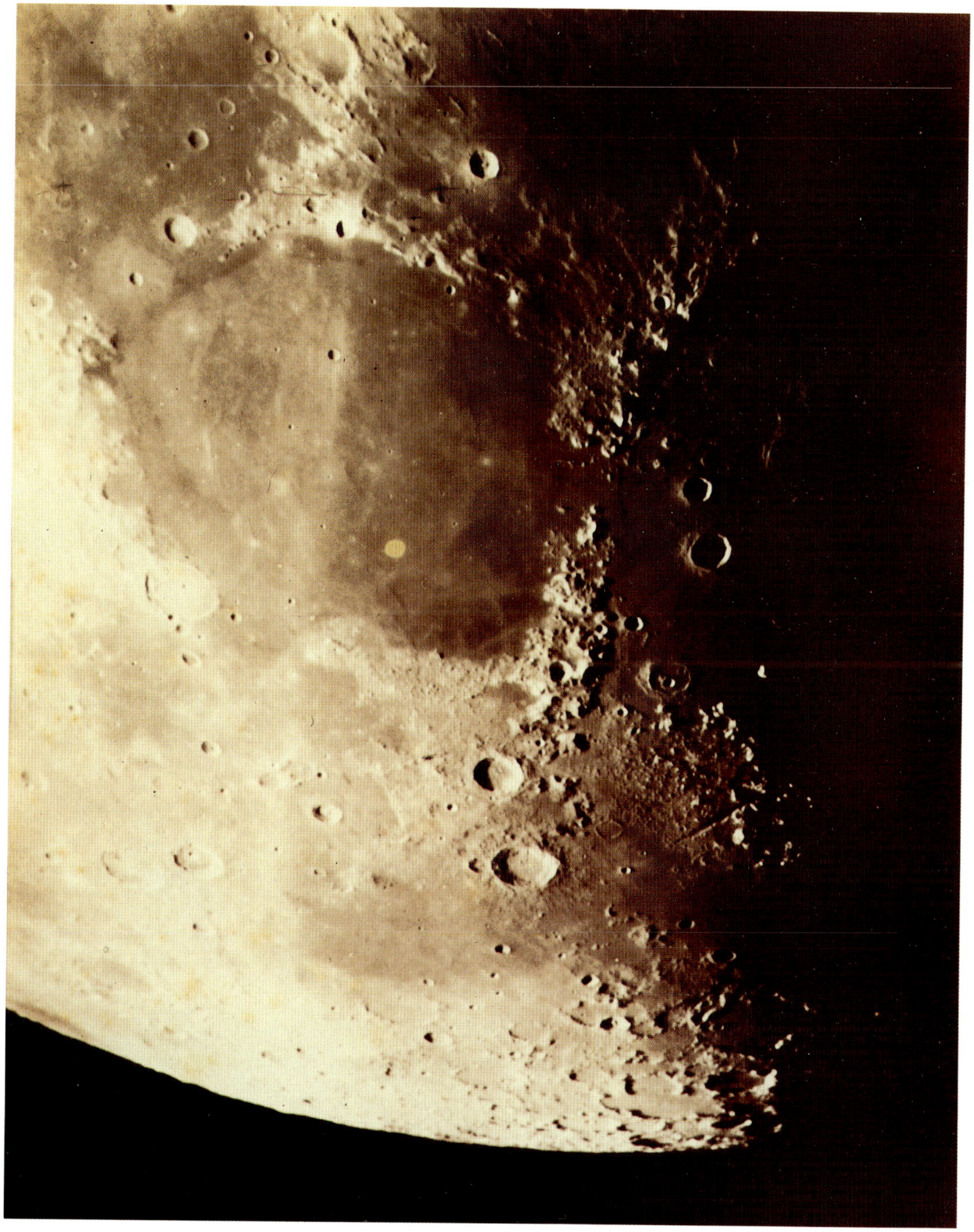

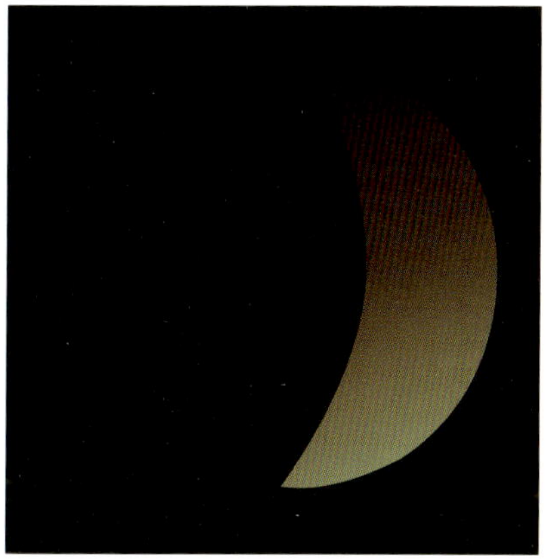

First Flights *of the* Twentieth Century

The Wright Brothers

On December 17, 1903, Orville and Wilbur Wright, known familiarly as the Wright brothers, achieved the first powered, sustained, and controlled airplane flight. By 1905 they had built the world's first practical engine-powered airplane. By 1910, there were airplane producers in France, England, Germany, and Russia, as well as the Wright brothers' company and others in the United States. Although commercial air transportation did not gain acceptance for another decade, the idea of human flight had captured the public's imagination.

Jacques-Henri Lartigue

Jacques-Henri Lartigue began to take photographs at the tender age of five, and in the years before World War I, he captured with a wonderful freshness the zest for life enjoyed by his prosperous family in France. He also kept a diary. The following passage, written when he was sixteen, accompanies his own photographs of his brother Zissou's efforts to fly in a glider.

> A construction center for gliders is set up in the park at Rouzat [France], near the big winery. Pirou, the village carpenter, and several friends have come to help [my brother] Zissou. Armed with wood, saw, wire, nails and sheets (which Zissou sneaked out of Maman's linen closet) they begin to work. When all the different parts of the big glider are finished, everybody — the maids, the butler, the nurses, the chauffeur, the cook and even the farmer — come to help assemble the machine. It's beautiful! The model is based on that of the Wright brothers' glider.
>
> On the first day the wind is strong enough,... we take the glider out on the field and Zissou perches himself inside it, ready to take off.... He starts walking. He runs. Louis and the farmer pull the cords as hard as they can. One of Zissou's feet goes up in the air ... then up shoots the other one. Zissou is airborne! But ... the glider starts cracking! Zissou falls. The ZYX 24 is broken, but never mind ... Zissou has flown.[2]

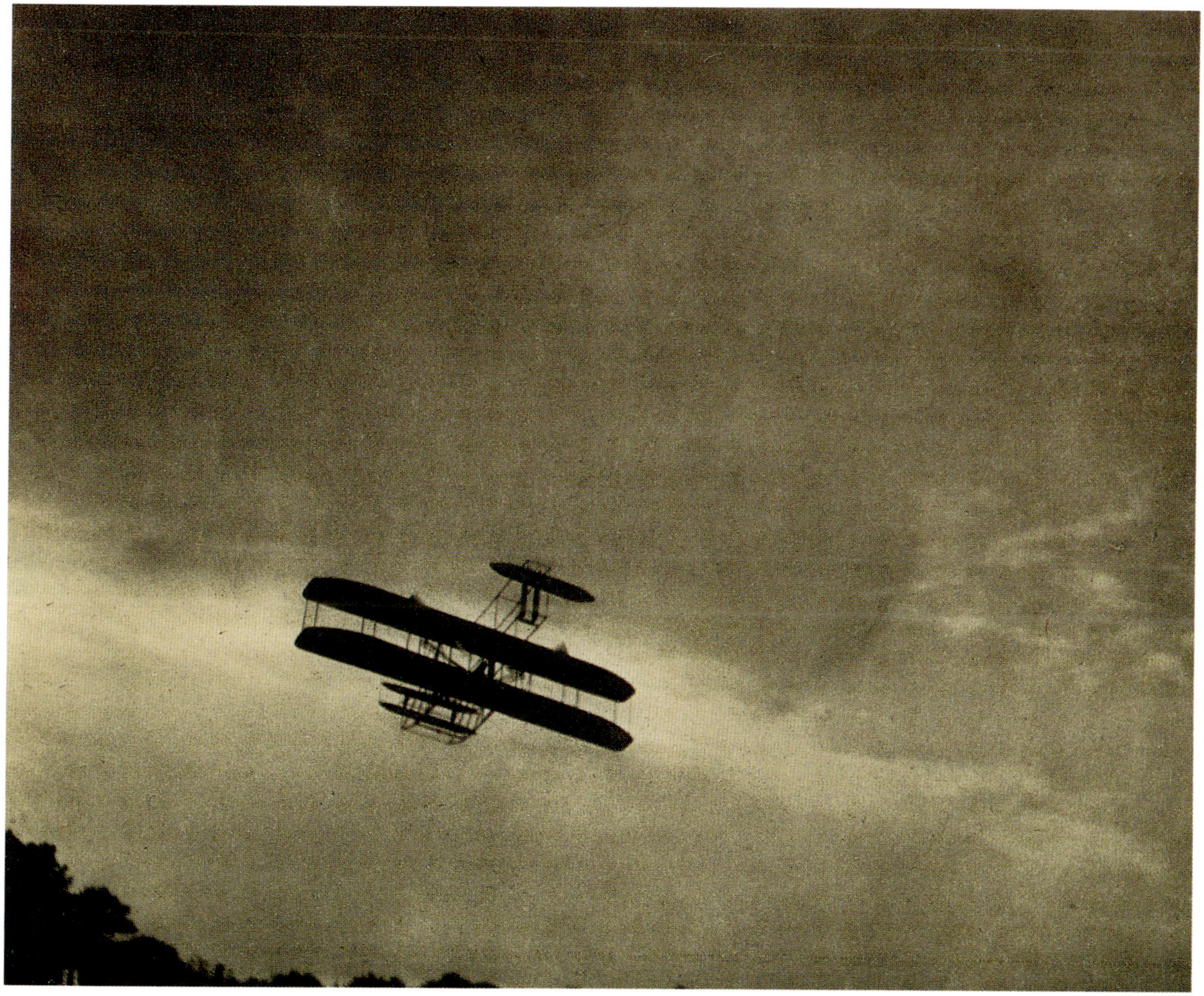

Alfred Stieglitz. *The Aeroplane*, from *Camera Work* (No. 36). Published 1911. Photogravure on tissue.
Museum purchase with funds provided by Target Stores.

THE FIRST MEN IN SPACE

Soviet cosmonaut Yuri A. Gagarin became the first man to orbit the Earth on April 12, 1961; his flight lasted one hour and forty-eight minutes and made one complete orbit. On May 5 of this same year, the United States launched Alan B. Shepard, Jr. in a successful suborbital flight, followed by Virgil I. "Gus" Grissom's suborbital flight in July. In August, Russian cosmonaut Gherman S. Titov flew seventeen orbits and spent a day in space aboard Vostok 2. On February 20, 1962, John H. Glenn, Jr. became the first American to orbit the Earth.

None of the astronauts was prepared for the tremendous public adulation or worldwide media attention they received. More than 500,000 people flocked to Cape Canaveral, Florida, for Shepard's launch; before Glenn's orbit on February 20, some people had been camping on the beaches since January 23, when the flight was originally scheduled. After the first astronauts returned, they received medals from President John F. Kennedy and were honored at ticker-tape parades in several cities. People sought their autographs; the press aggressively pursued them and their families. Hundreds of thousands of people from around the world wrote to them. For the crews, the response was both heartwarming and unnerving.

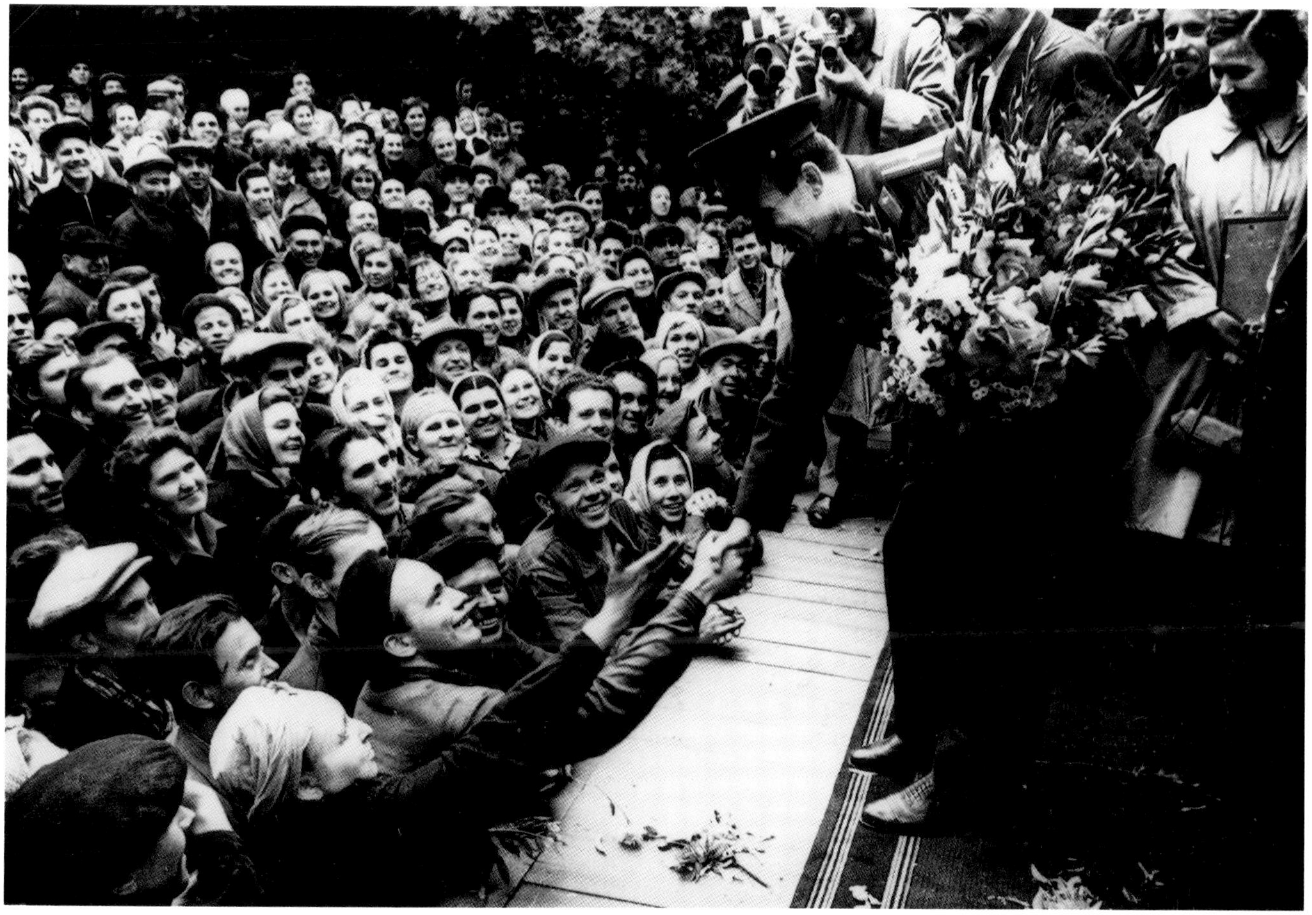

Rafail Diament. *Soviet Cosmonaut Yuri Gagarin, "the first man in space," after his successful orbital flight on April 12, 1961.* 1960s. Gelatin silver photograph. Courtesy of Robert Koch Gallery, San Francisco.

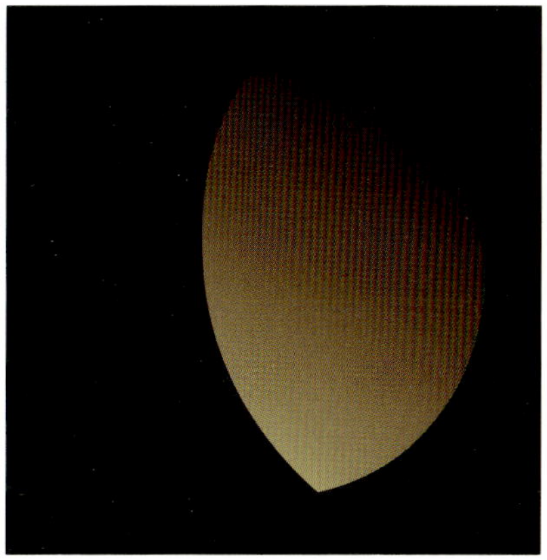

The NASA Program

President Kennedy's Challenge

The Soviet Union's Sputnik, launched in October 1957, and Yuri Gagarin's orbit in April 1961, forced the United States to undertake major adventures in space. As Russia's chief rocket engineer, Sergei Korolev, had announced after launching Sputnik, "Today, the dreams of the best sons of mankind have come true. The assault on space has begun."[3]

The "space race" became a fateful test and presage of the Cold War conflict between the superpowers. During President Dwight D. Eisenhower's administration, the National Aeronautics and Space Administration (NASA) was established, and the first American manned space flights were planned. After President John F. Kennedy's inauguration, Alan B. Shepard, Jr. became the first American to fly in space in May 1961, followed by Grissom and Glenn.

President Kennedy escalated the assault to compete in space on May 25, 1961, when he recommended to Congress that the United States should achieve manned exploration of the moon before the end of the decade. "No single space project," said Kennedy, "will be more impressive to mankind or more important for the long-range exploration of space; and none will be so difficult or expensive to accomplish."[4] Kennedy's challenge was met on July 20, 1969, with less than six months left in the decade.

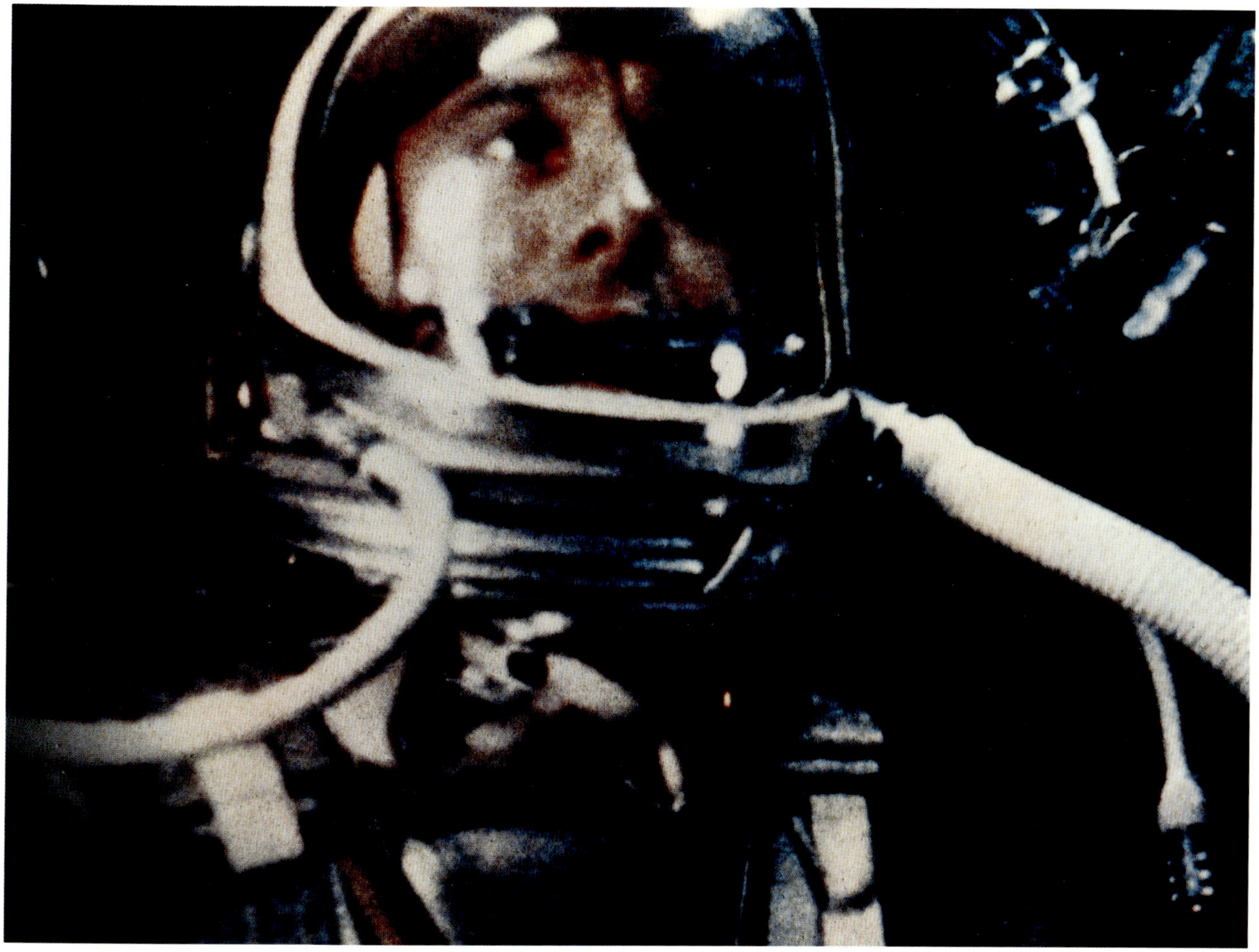

Mission: Mercury-Redstone 3, Freedom 7, May 5, 1961. *Alan B. Shepard, Jr., the first American in space.*
Dye-transfer photograph. The Museum of Fine Arts, Houston; The Target Collection of American Photography;
museum purchase with funds provided by Target Stores.

A Chimp Named Ham

Insisting that there were too many unknowns about space flight, especially weightlessness and what could be unexpectedly high g-forces, NASA's medical teams and psychologists decided to send a chimpanzee in a Mercury-Redstone flight before they sent a human being. The astronauts, especially Alan Shepard, thought it was a huge waste of money, time, and manpower.

On January 31, 1961, a chimp named "Ham" experienced a rough ride on a not-quite-perfect rocket. It climbed at a slightly higher angle than planned, driving Ham back into his couch with a g-force of seventeen, i.e., seventeen times his own weight, much more than anticipated. The on-board equipment failed, the electrical system and light tests went haywire, and the rocket overshot its landing site by one hundred miles, although Ham survived. Yet another Mercury-Redstone test flight was launched—this time a perfect one—before Alan Shepard successfully rode a rocket through the atmosphere.

Mission: Mercury-Redstone, January 31, 1961. *Chimpanzee Ham being prepared for a successful suborbital flight in a Mercury capsule.* Dye-transfer photograph. The Museum of Fine Arts, Houston; The Target Collection of American Photography; museum purchase with funds provided by Target Stores.

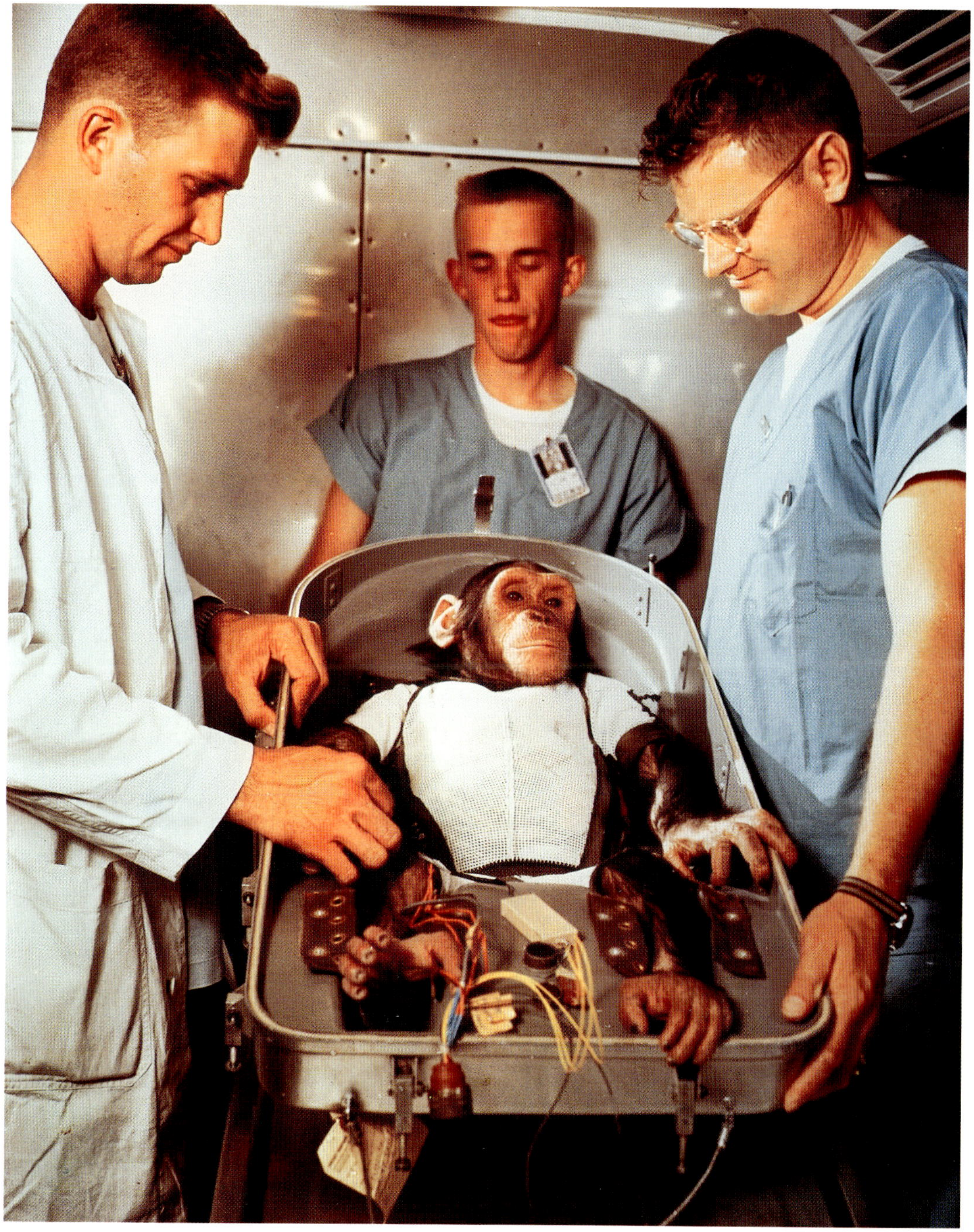

Florida, Texas, and California

There are three primary sites of NASA explorations: the launch facilities at John F. Kennedy Space Center at Cape Canaveral, Florida; the Lyndon Baines Johnson Space Center in Houston; and Edwards Air Force Base in California.

The choice of Florida was eerily prophesied by the French novelist Jules Verne in *From the Earth to the Moon and All Around the Moon,* published in two parts in the United States, beginning in 1865. In the novel, an aluminum projectile carrying three men and two dogs as passengers is fired from a point in Florida, not far from what is now the Kennedy Space Center. The manned projectile orbits the moon, then fires rockets which send it back to Earth, where it splashes down in the Pacific to be retrieved by a U. S. naval vessel. In fact, after World War II, the U.S. Army built a missile test range at Cape Canaveral. Located mid-peninsula on the Atlantic Coast, the Cape became the natural site to locate the nascent American space effort. After the Apollo program was initiated, NASA purchased acreage on adjoining Merritt Island.

The choice of Houston as the home of NASA's mission control center can be attributed to the combined political clout of then-Vice President Lyndon B. Johnson and Representative Albert Thomas, a key member of the House Appropriations Subcommittee. Although Houston had the required advantages of highway, rail, water, and air transportation, and had been on the short-list of sites for the Apollo program, the city could not have secured the winning bid for NASA's Manned Spacecraft Center (later Johnson Space Center) without the critical lobbying of Johnson and Thomas. In Texas, the American frontier of the nineteenth century—the legendary Wild West—merged with the American frontier of the twentieth century—Outer Space, which previously existed only in our imagination. The first operational flight controlled from the new center was Gemini 4 in 1964.

Edwards Air Force Base is a military installation on the edge of California's Mojave Desert. The flat, desert lake bed is favored by test pilots because it stretches for miles in all directions, offering the world's longest runways. The military base is sparsely populated, thus people are rarely disturbed by the sound of crashing sonic booms. In 1958 NASA based its search for America's first astronauts on an applicant pool of test pilots at Edwards. For the same reasons that test pilots prefer Edwards's landing fields, the base was chosen as the initial landing site for the space shuttle—the first spacecraft designed to carry its crew back to a runway landing.

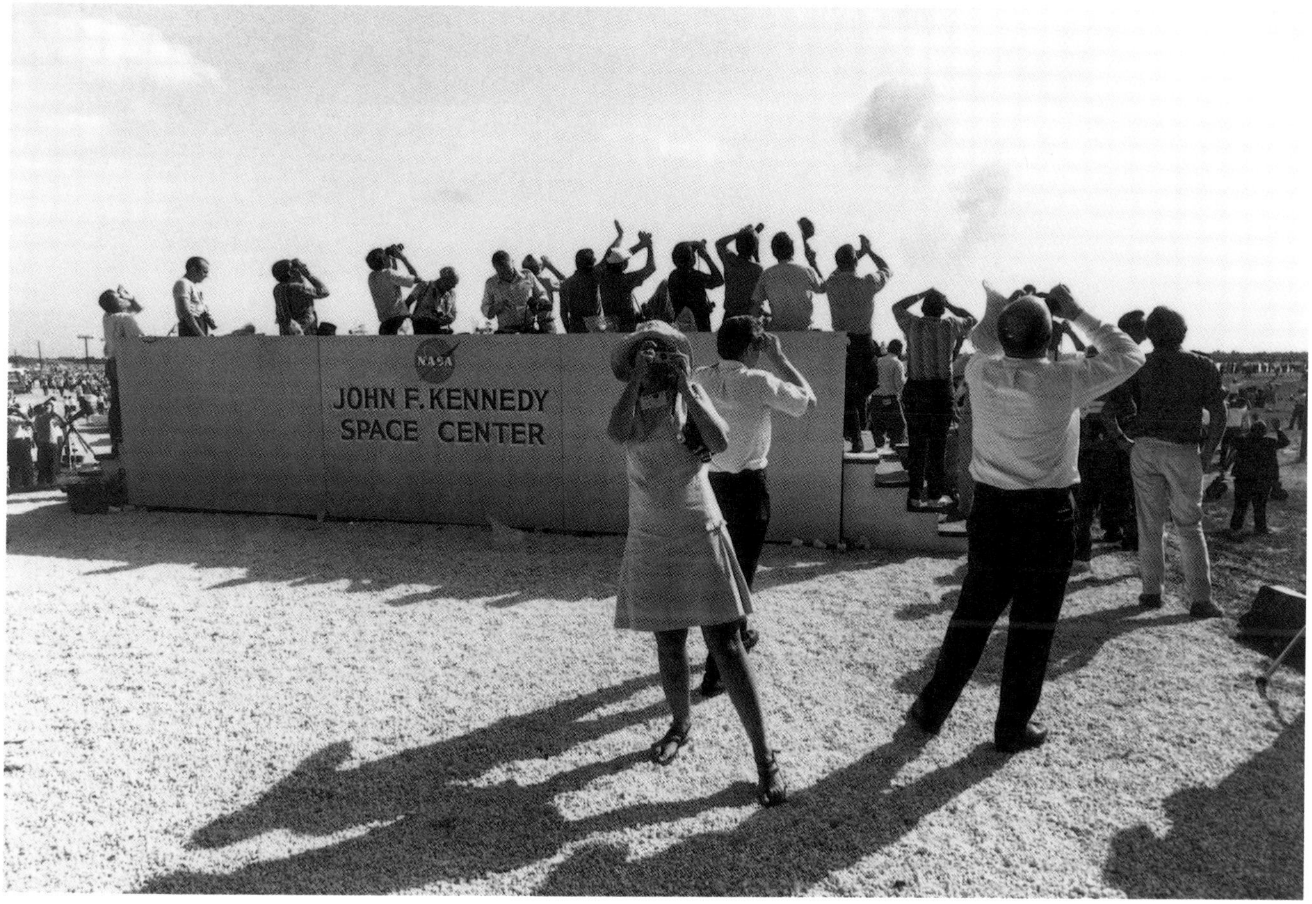

Garry Winogrand. *Apollo 11 Moon Shot, Cape Canaveral, Florida.* 1969.
Gelatin silver photograph. The Museum of Fine Arts, Houston; The Allan Chasanoff Photographic Collection.

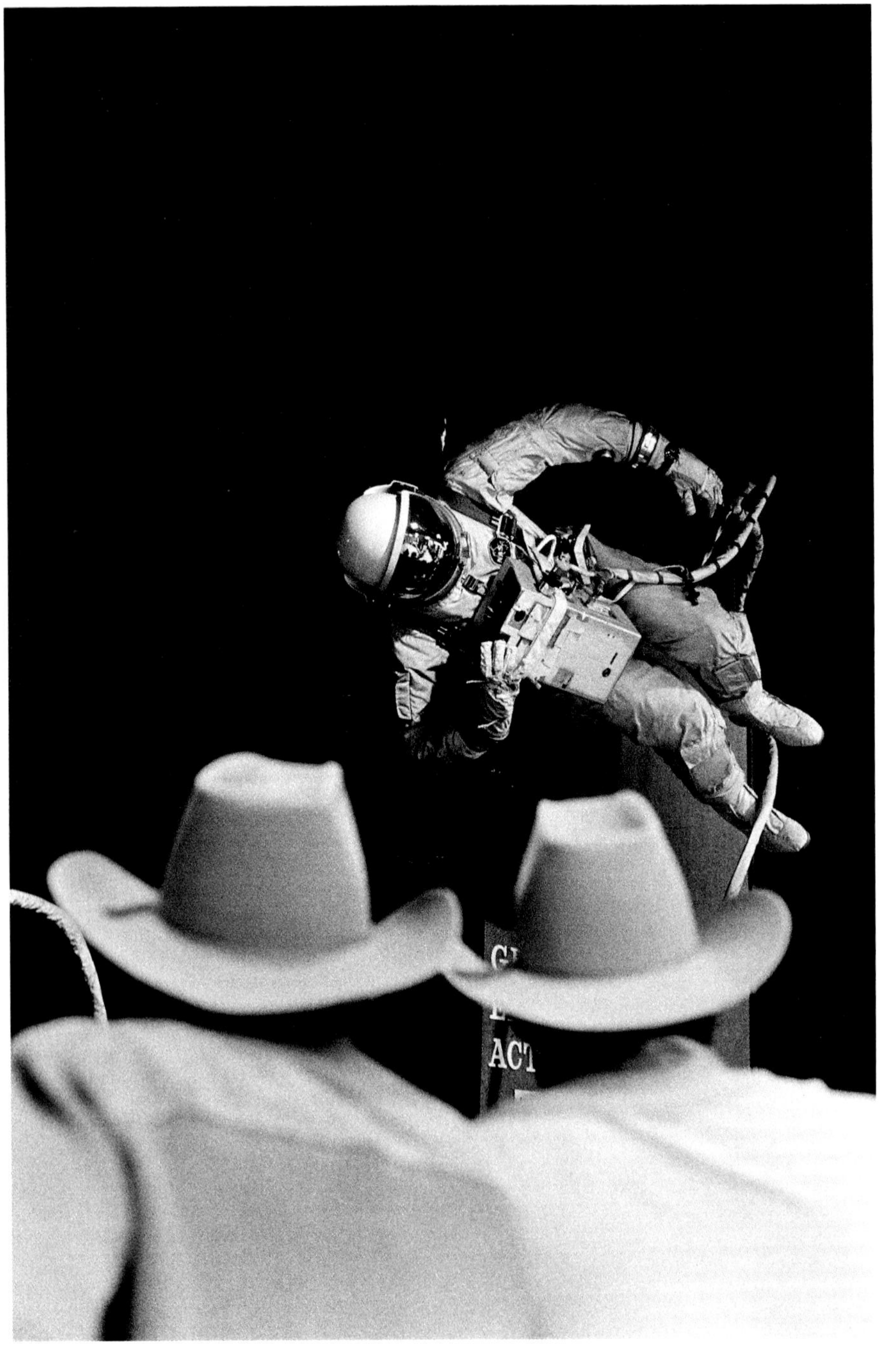

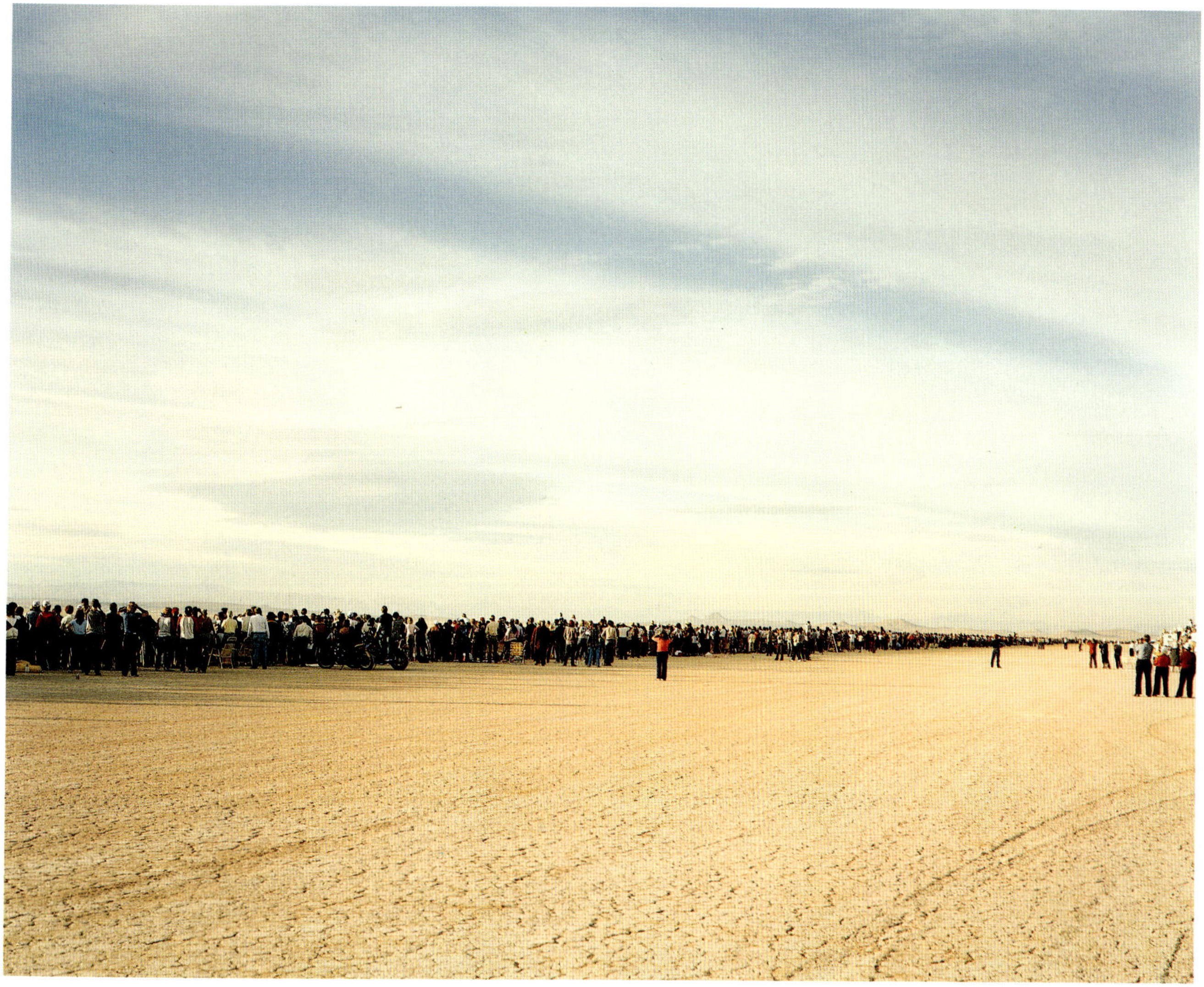

Above: Richard Misrach. *The Shuttle Landing, Edwards Air Force Base.* 1983. From the series *Desert Canto II: The Event I.* Color coupler photograph. Courtesy of Fraenkel Gallery, San Francisco; Robert Mann Gallery, New York; and the photographer.

Opposite: Ikko Narahara. *Houstonscapes #28-1.* 1985. Gelatin silver photograph. The Museum of Fine Arts, Houston; gift of FotoFest and the artist.

GEMINI

Gemini, the heavenly twins of the zodiac and the constellations, was chosen as the name for the second stage of NASA's program because the capsule employed two men. Much more sophisticated than the Mercury capsule, the Gemini capsule was capable of rendezvous and docking, of changing its orbit as well as its orientation, and of opening a hatch to allow one of its passengers to "walk" in space.

The Gemini program became the indispensable bridge between the Mercury and Apollo ventures. Once NASA had chosen "Lunar Orbital Rendezvous" as its method for landing on the moon, each of the necessary procedures and equipment had to be formulated and tested first in Earth's orbit. NASA also needed to test how the astronauts would respond to a week or more of weightlessness and to test their space suits. Temperatures outside the craft range from 250 degrees when directly exposed to solar radiation to 150 degrees below zero when floating in the spacecraft's shadow. The suit not only must protect the astronaut, keep his or her body's temperature at correct levels, and provide oxygen, but also allow enough flexibility for the astronaut to perform complicated tasks.

James A. McDivitt, Mission: Gemini-Titan IV, June 3-7, 1965. *Edward H. White performs the United States's first space walk which NASA refers to as an extravehicular activity (E.V.A.). In his hand is a pressurized gas gun to propel his maneuvers. The gold-covered umbilical tether connects his suit with the life support system of the space craft.* Dye-transfer photograph. The Museum of Fine Arts, Houston; The Target Collection of American Photography; museum purchase with funds provided by Target Stores.

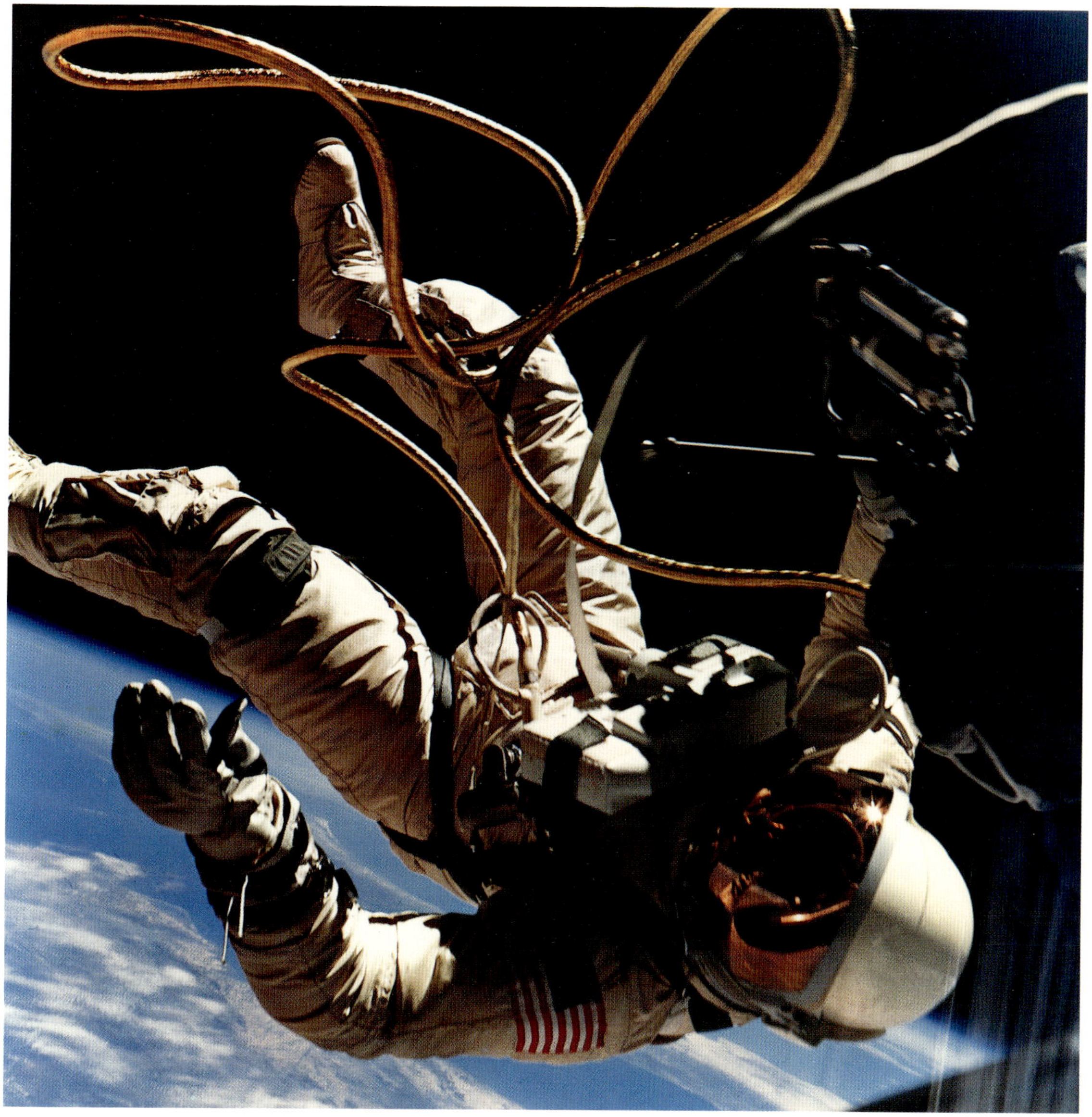

Photographs of the Moon from Unmanned Spacecrafts

The first photographs of the moon from a spacecraft were made by the USSR Luna III at extremely long range on October 3, 1959. American lunar photography was planned as three flight projects called Ranger, an impact (crash) craft; Orbiter, a moon-orbiting craft; and Surveyor, a soft-landing craft. Thousands of photographs were taken either by traditional photographic systems or by television cameras, which were modified to work in space. All the pictures were electronically transmitted to Earth. The Orbiter spacecraft obtained high-resolution information of the lunar surface that proved essential to selecting landing sites for the Apollo missions as well as to studying the moon's surface.

Landing on the Moon

At 4:17 p.m. EST, July 20, 1969, Neil A. Armstrong informed mission control: "Houston, Tranquility Base here. The Eagle has landed." Eight years after President Kennedy's challenge, the crew of Apollo 11's lunar module, Eagle, climbed down its ladder to the moon's surface. Both Armstrong and Edwin E. "Buzz" Aldrin, Jr. discovered that being on the moon and moving in one-sixth gravity was an exhilarating experience. Aldrin described himself as "buoyant and full of goose pimples." Because the moon's surface is fine and powdery, both Aldrin and Armstrong were able to leave clear footprints, which Armstrong recorded as unprecedented scientific and historic information.

Neil A. Armstrong, Mission: Apollo-Saturn 11, July 16-24, 1969. *Edwin E. "Buzz" Aldrin, Jr. on the moon. Because Neil Armstrong was the photographer throughout the moon landing, the only pictures of Armstrong on the moon are his image reflected on the visor of Aldrin's suit helmet.* Dye-transfer photograph. The Museum of Fine Arts, Houston; The Target Collection of American Photography; museum purchase with funds provided by Target Stores.

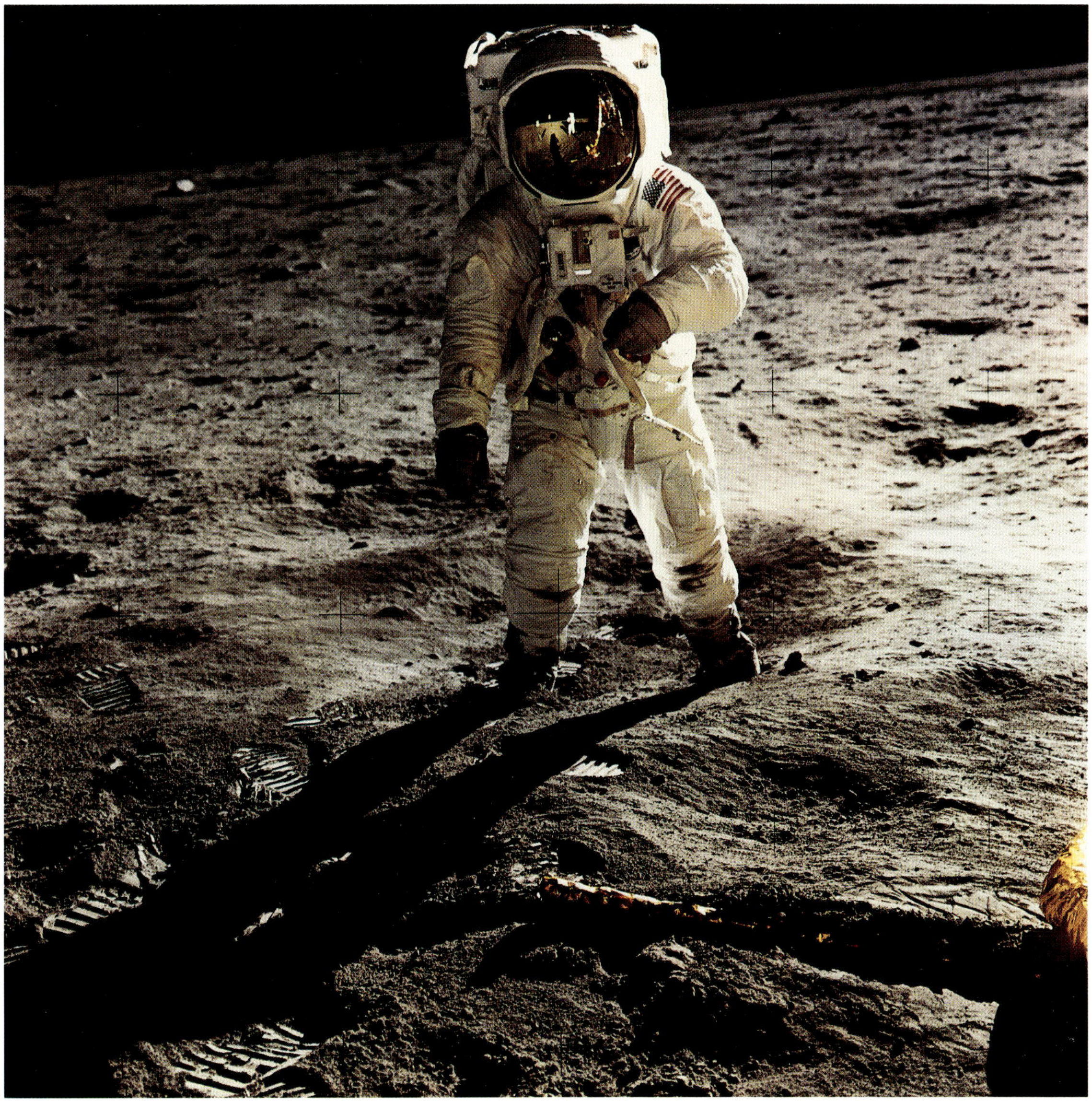

Revelations

Typical of many professionals working in the hard-core sciences, astronauts generally do not add philosophical import to their tasks. Because they are often uncomfortable with media scrutiny and the public demands of their profession, they respond briefly to questions that do not require factual answers. Also, their schedules in space are usually so crowded with duties that they do not have much time for reflection, let alone "sightseeing." The three consistent exceptions to the tone of their customary responses have been their attempts to describe the experience of weightlessness and of looking back at the planet Earth from the moon, or, when orbiting the Earth, of marveling at its great expanses.

Alan Shepard was the first to exult, "I'm free!," when he first experienced weightlessness. Similarly, when speaking of his partner Alexei A. Leonov, who had just returned from the first space "walk," Cosmonaut Pavel I. Belyaev said, "He looks like a man who has just been reborn, a man who has just come back from another world."[5] Edgar Mitchell said that walking on the moon was a "peak experience" for him, in which the "presence of divinity became almost palpable."[6]

When describing their unique vistas of Earth, astronauts tend to be uncharacteristically expansive. Upon seeing the Earth "rise" above the lunar horizon, Frank Borman (Apollo 8) told mission control in Houston, and the listening world, "This is the most beautiful, heart-catching sight of my life." Joseph P. Allen (Shuttles 5 and 51-A) described the ever-changing scenes of the Earth outside his orbit craft as "endlessly fulfilling."[7]

Michael Collins, Mission: Apollo-Saturn 11, July 16-24, 1969.
Lunar module, Eagle, returns to command/service module, Columbia, after the first lunar landing. Dye-transfer photograph. The Museum of Fine Arts, Houston; The Target Collection of American Photography; museum purchase with funds provided by Target Stores.

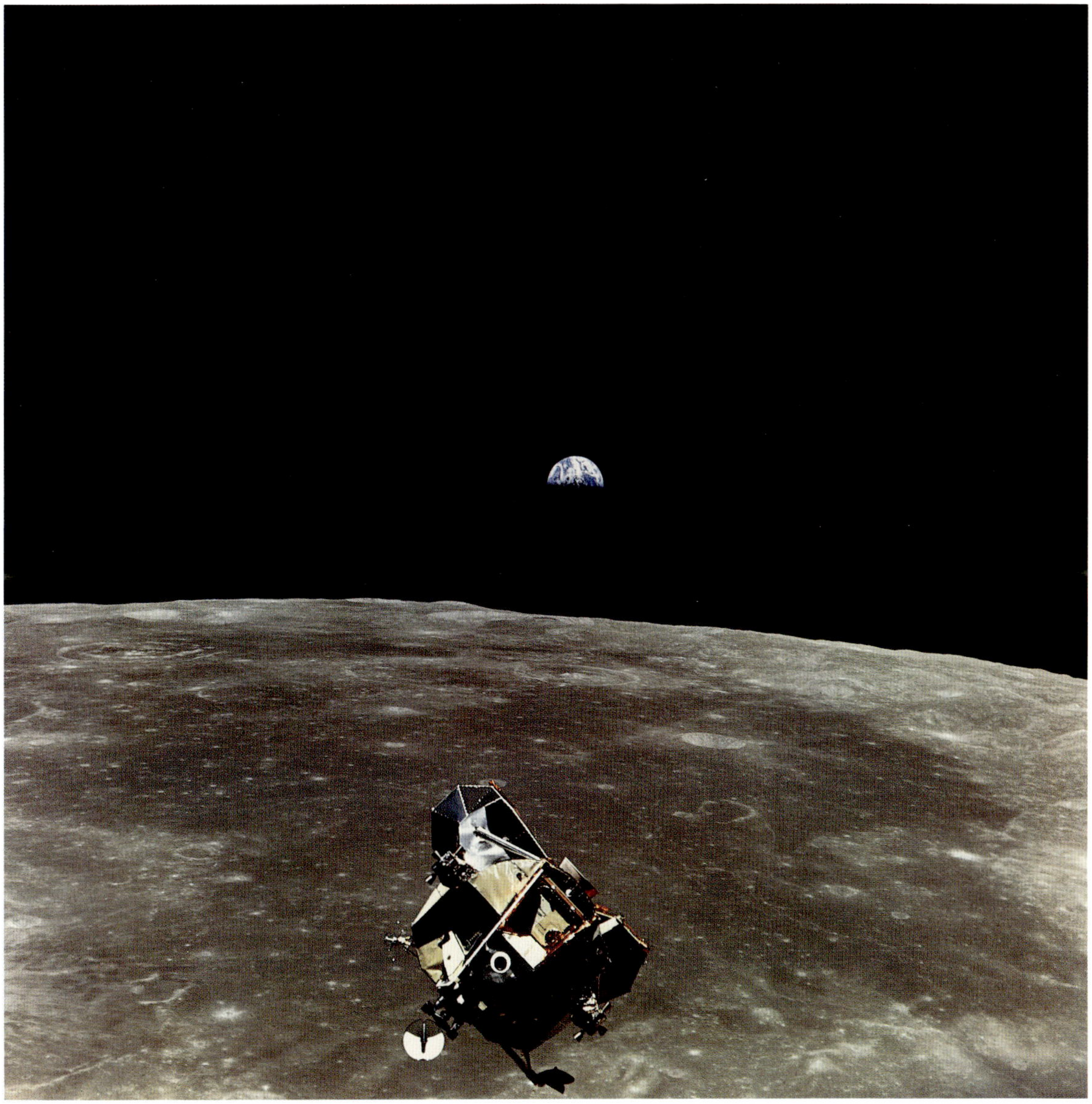

The Whole Earth

The spectacular view of the whole planet Earth that the Apollo astronauts captured in photographs is credited with giving tremendous momentum to the then-nascent environmental movement. The power of seeing the Earth as one planet, rather than as separate nations and regions, has been cited again and again by the returning astronauts as among their most unforgettable experiences. William Anders (Apollo 8) said, "It was this view of the fragility and finiteness of the Earth that is the impression, frankly, that I hold more in my head than any other."[8] Alan Shepard (Apollo 14) wrote that he understood how "the Earth seemed almost limitless to its people with its vast oceans and upheaving mountains, where there was always a distant horizon and changing dawns and sunsets. But from here, from the moon, it is, in fact, very finite, very fragile ... so incredibly fragile. That thin, thin atmosphere, the thinnest shell of air hugging the world—it can be blown away so easily! A meteor, a cataclysmic volcano, man's own uncaring outpourings of poison..."[9]

Even without seeing the whole Earth, and only the great expanses visible to orbiting astronauts, Byron K. Lichtenberg (Shuttle 9) acknowledged, "Your perspective really changes.... You don't see any of the borders, you don't see any of the cultures, you don't see the different languages.... So there's the sense of no difference, just a subtle gentle blending from one region to the next."[10]

Mission: Space Shuttle 40, Columbia, June 5-14, 1991. *Iran, Kuwait Fires, Persian Gulf.* Dye-transfer photograph. The Museum of Fine Arts, Houston; The Target Collection of American Photography; museum purchase with funds provided by Target Stores.

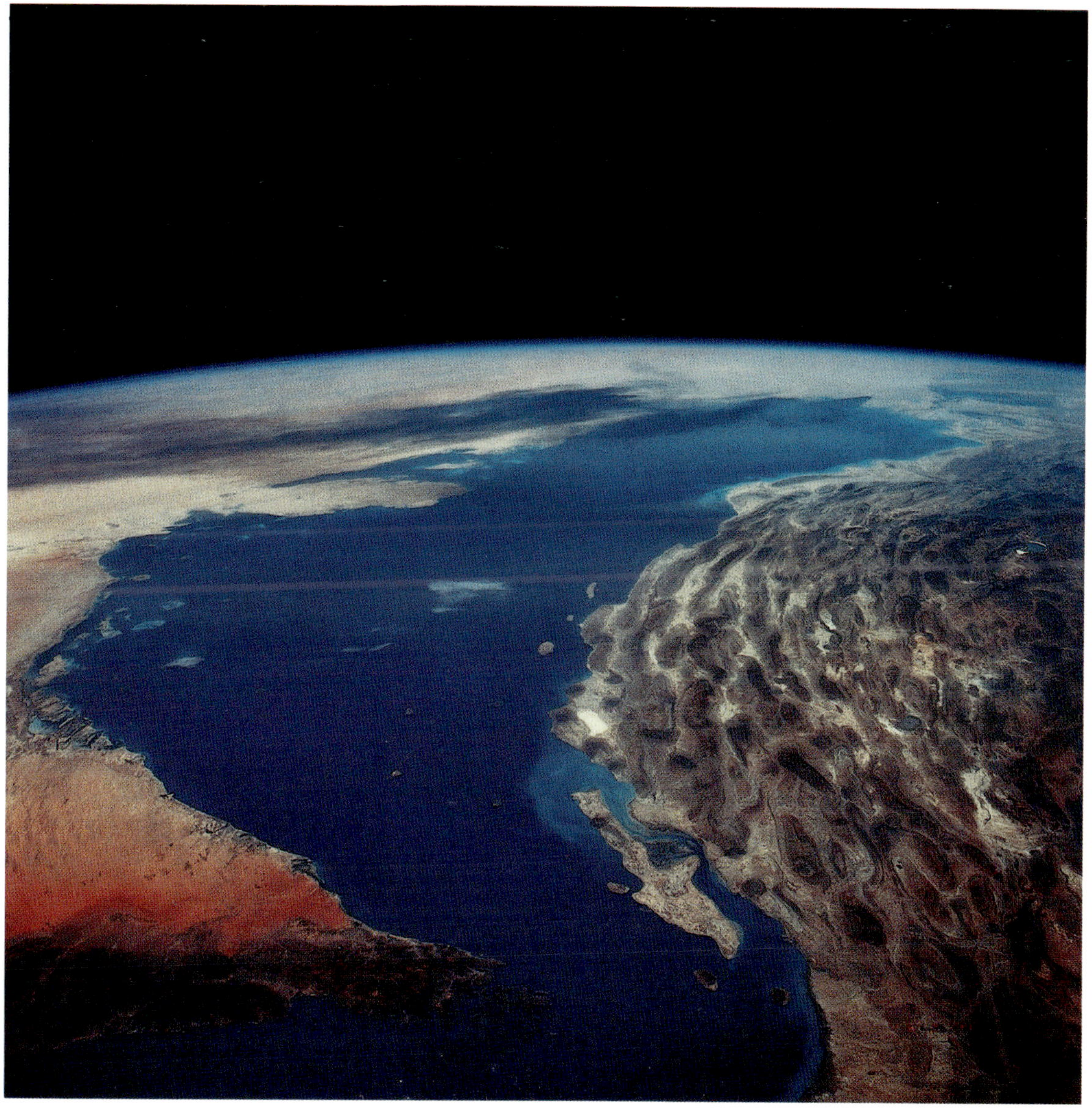

Space Shuttle

The United States's space shuttle is the world's first reusable spacecraft, designed to fly like a rocket in space and like a glider in the atmosphere. Once the shuttle is in orbit, the astronauts can open the doors of the payload bay to perform experiments and tasks, such as retrieving, repairing, and/or launching satellites. A fifty-foot-long remote manipulator arm assists in payload procedures.

The approximate size of a DC-9 airliner, the shuttle can carry a crew of nine as well as eighteen tons of cargo. Shuttle missions can last seven to ten days. The shuttle is also the first spacecraft designed to carry its crew back to a runway landing. If possible, shuttles land at Kennedy Space Center in Florida, but Edwards Air Force Base is also available, as are other bases worldwide. Then the shuttle is ferried to Florida on the back of a Boeing 747.

Mission: Space Shuttle 41-B, Challenger, February 3-11, 1984. *Bruce McCandless making the first untethered space walk with the manned maneuvering unit (M.M.U.).* Dye-transfer photograph. The Museum of Fine Arts, Houston; The Target Collection of American Photography; museum purchase with funds provided by Target Stores.

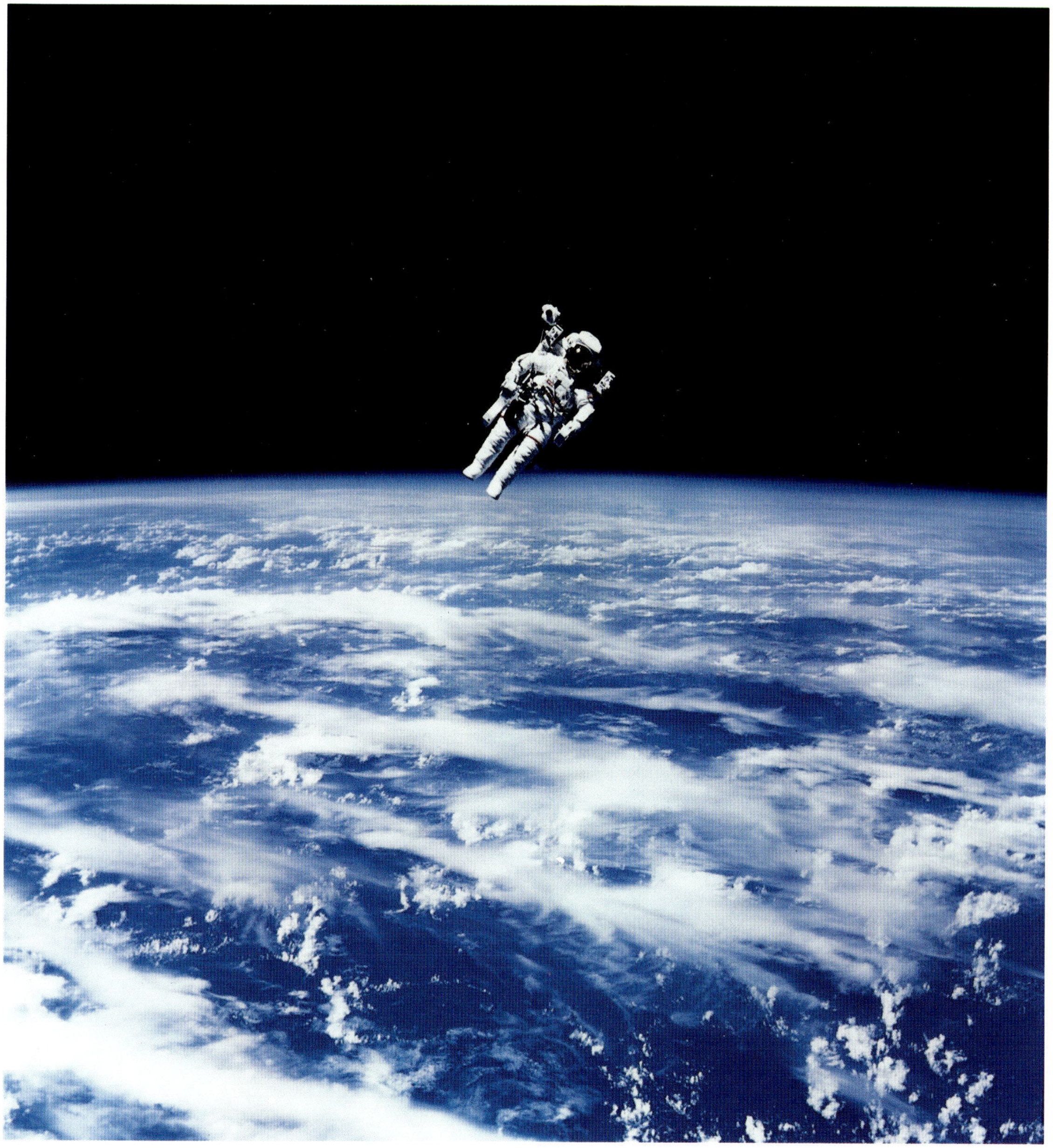

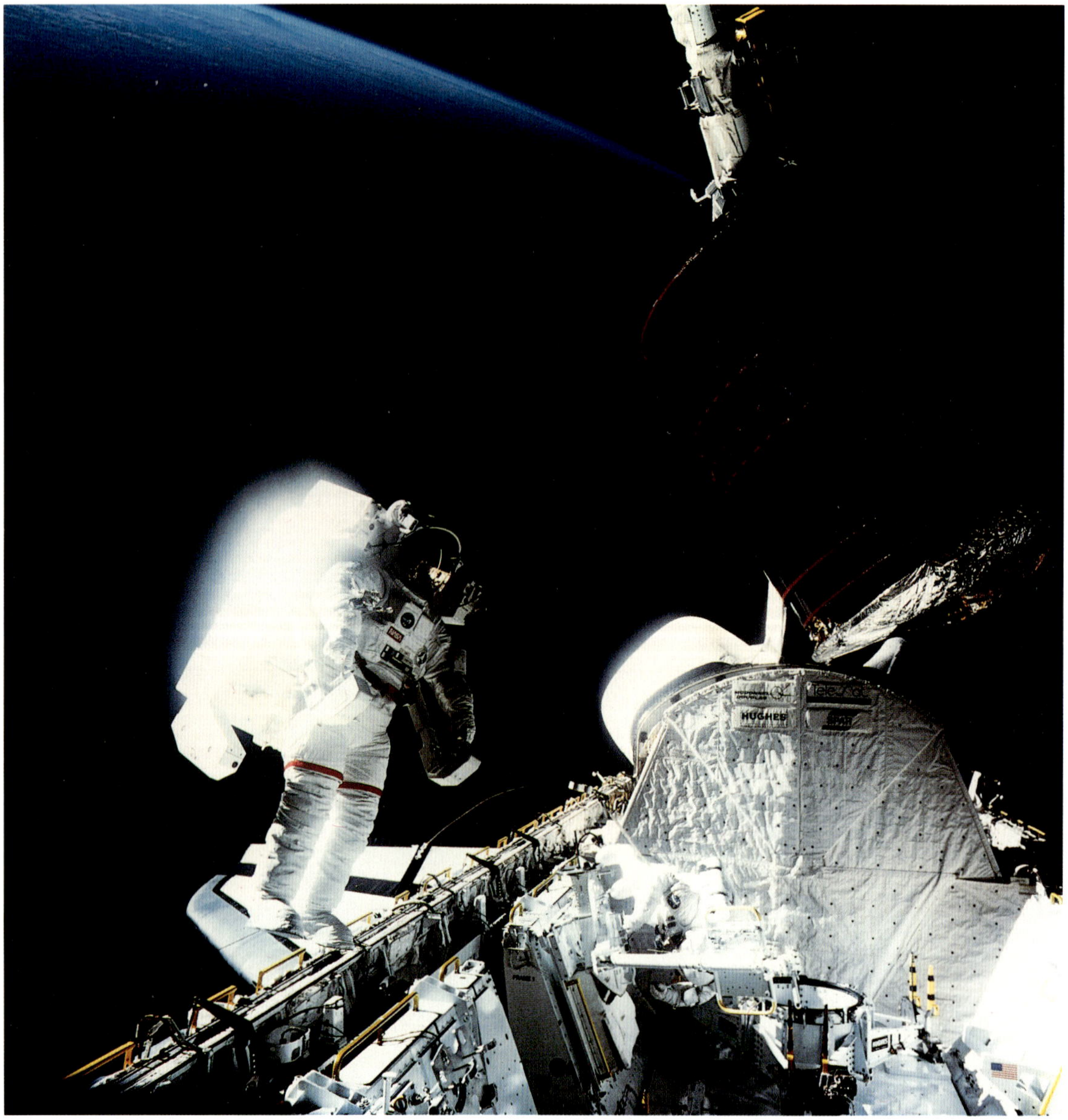

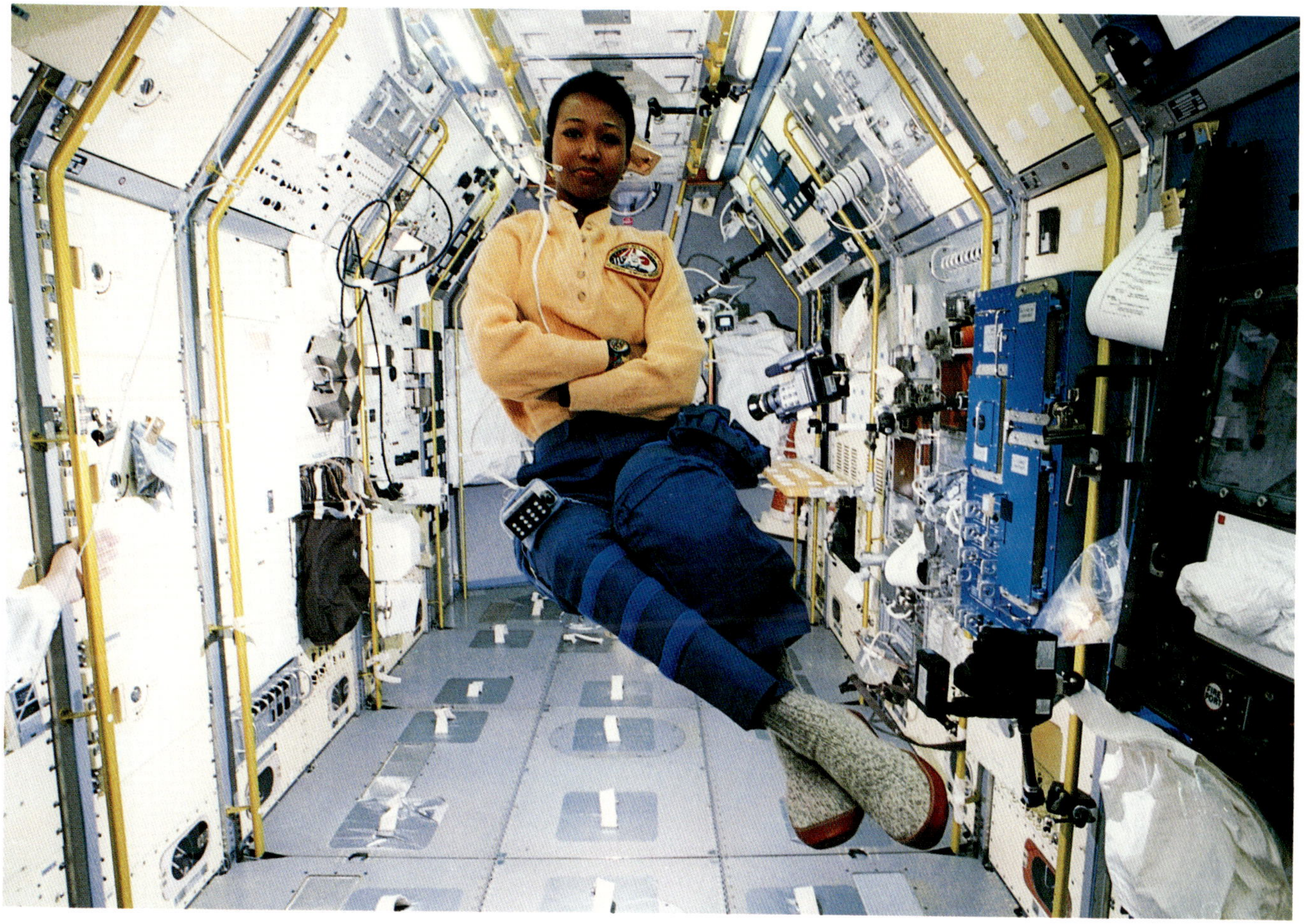

Above: Mission: Space Shuttle 47, Endeavour, September 12-20, 1992. *Dr. Mae C. Jemison, first black female astronaut.* Dye-transfer photograph. The Museum of Fine Arts, Houston; The Target Collection of American Photography; museum purchase with funds provided by Target Stores.

Opposite: Mission: Space Shuttle 51-A, Discovery, November 8-16, 1984. *Joseph P. Allen maneuvers near the cargo bay as Dale A. Gardner works in the bay. This mission was the first to use the remote manipulator system, otherwise known as the "arm," to retrieve errant spacecraft.* Dye-transfer photograph. The Museum of Fine Arts, Houston; The Target Collection of American Photography; museum purchase with funds provided by Target Stores.

Photography in Space

Photographing in space presents unique problems. It is often difficult for an astronaut to frame a picture carefully, because unless his or her feet are locked onto the floor or his or her body is braced, the astronaut-photographer will float away from the desired composition. There are also extreme and rapid shifts in light levels. On the sunlit side of the craft, the reflections are brilliant, and on the shadow side, the lack of light is absolute. When looking out over the space shuttle's payload bay, the light meter needle can gyrate 180 degrees with a slight shift in angle. Also, photographing from out of the spacecraft's windows yields reflections, refractions, and distortions that cannot always be avoided.

Among the advantages of space photography, however, are the spectacular views of the Earth and the moon, and the deep saturated colors, offset by brilliant reflections, that lend themselves to dramatic compositions.

The astronauts are supposed to make photographs of a purely scientific value, but their human experiences and observations appear on the picture discs from each flight. Charles Duke and Charles J. Precourt incorporated their families into their missions by arranging family portraits in the unique surroundings of space. Joseph P. Allen captured the experience of "parking" one's food in midair while eating. There are also documents of space high jinks, such as a picture of Elvis in a space suit, and a "For Sale" sign held aloft in space.

Charles J. Precourt, Mission: Space Shuttle 55, Columbia, April 26-May 6, 1993. *Snapshot of astronaut's family (Lynne, his wife, and three daughters: Michelle, Sarah, and Aimee).* Dye-transfer photograph. The Museum of Fine Arts, Houston; The Target Collection of American Photography; museum purchase with funds provided by Target Stores.

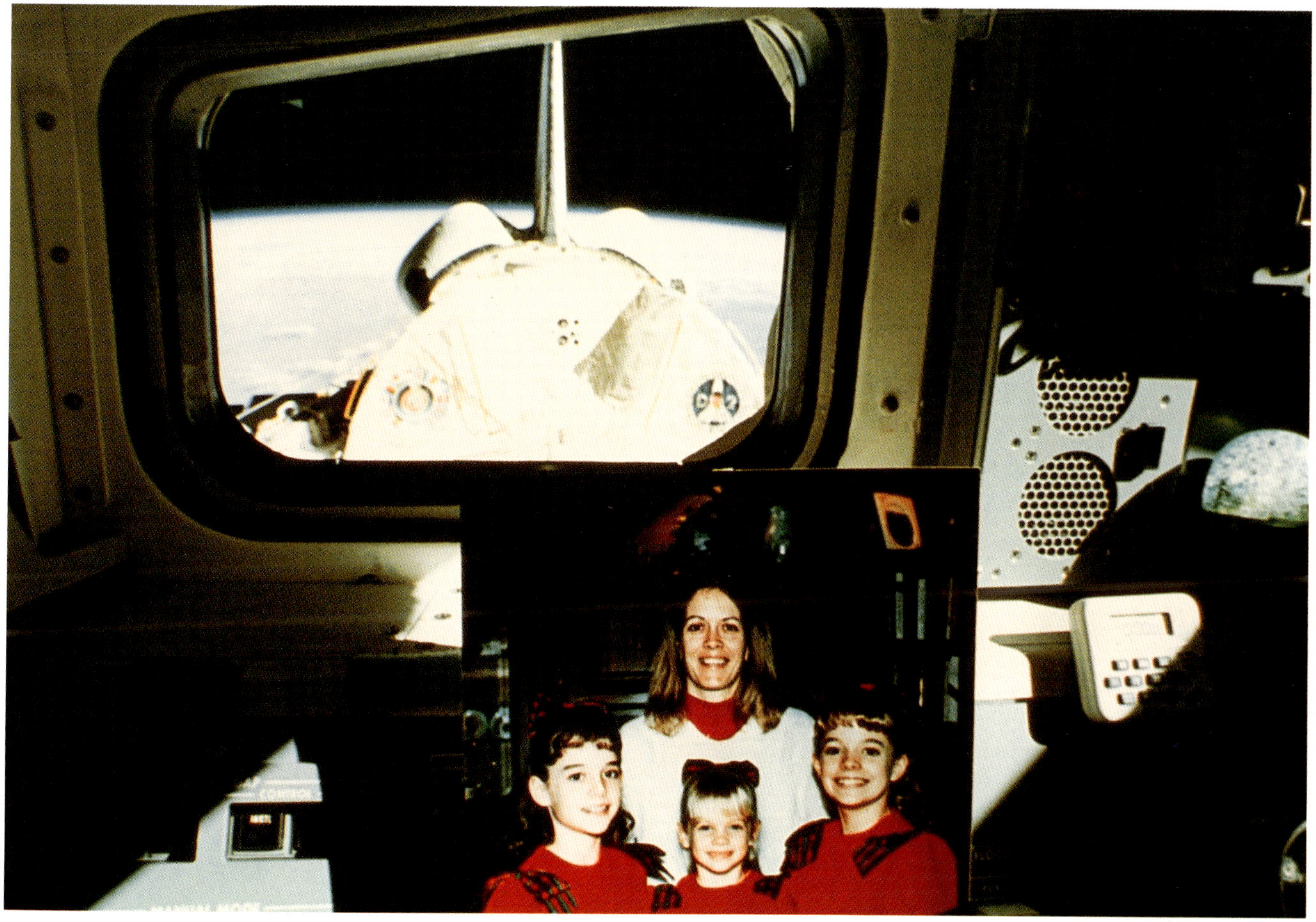

Repairing the Hubble Telescope

Space Shuttle 61 (December 2-13, 1993) was one of the most challenging and complex manned missions ever attempted. During a record five back-to-back spacewalks (totaling 35 hours, 28 minutes), two teams of astronauts completed the first servicing of the Hubble Space Telescope. Using the Remote Manipulator Arm, astronaut Claude Nicollier berthed the Hubble upright in the shuttle's payload bay. Then alternating teams—comprised of astronauts F. Story Musgrave, Jeffrey A. Hoffman, Kathryn C. Thornton, and Tom Akers—replaced malfunctioning parts of the telescope. During the mission, Akers set a new cumulative U.S. spacewalking record of 29 hours, 39 minutes, topping Eugene Cernan's 20-year-old record of 24 hours, 14 minutes. Thornton is the leading U.S. female spacewalker with a total of 21 hours, 10 minutes.

Mission: Space Shuttle 61, Endeavour, December 2-13, 1993. *Hubble Space Telescope with panels extended in the payload bay of Endeavour during HST repair mission.* Dye-transfer photograph. The Museum of Fine Arts, Houston; The Target Collection of American Photography; museum purchase with funds provided by Target Stores.

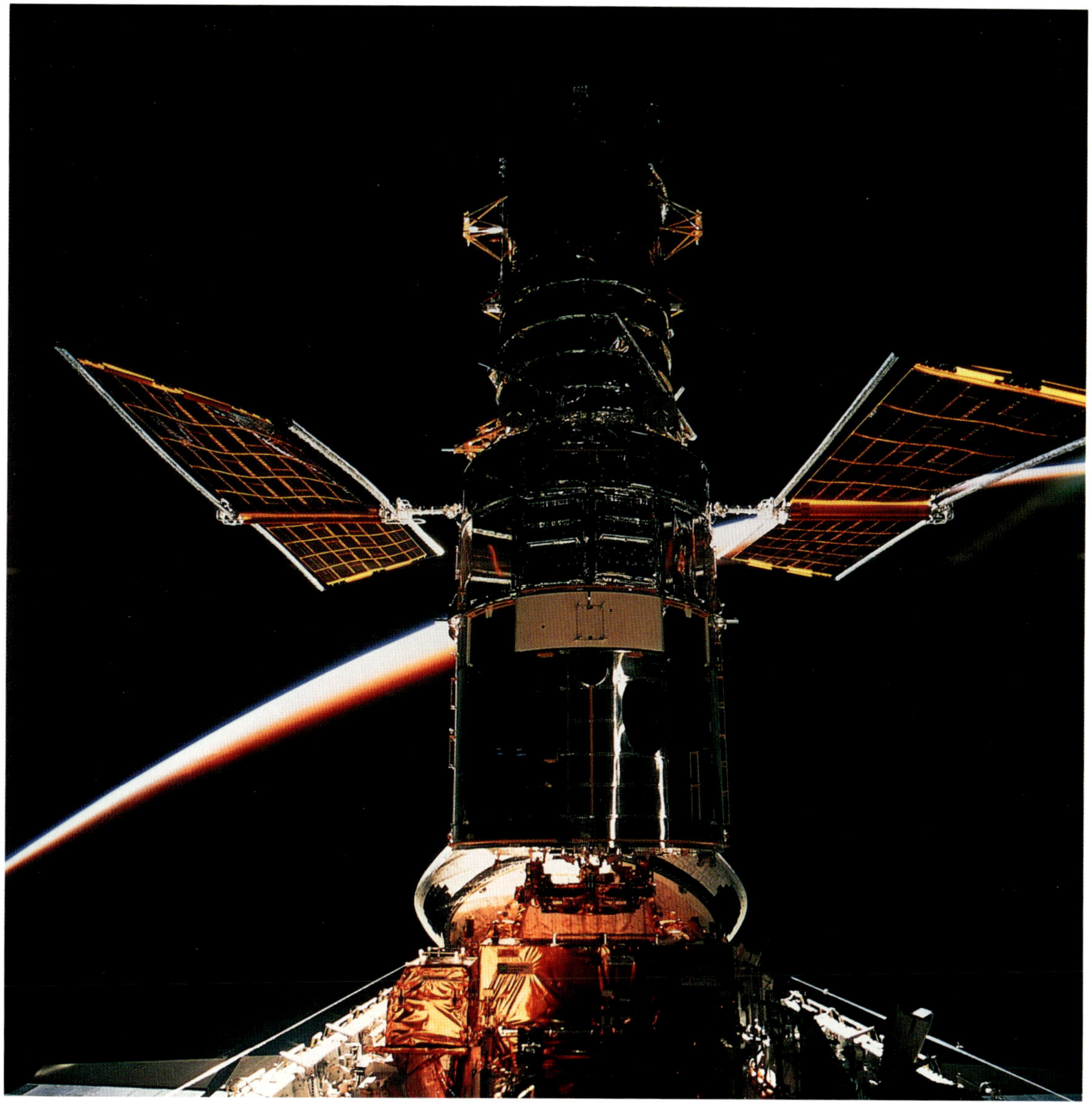

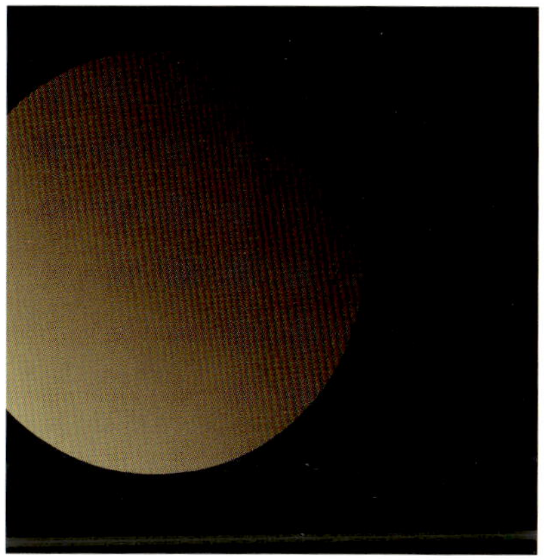

Flights *of the* Imagination

Dancing to Restore an Eclipsed Moon

When Edward Sheriff Curtis made this photograph in 1915, he wrote "The Qágyuhl people thought that an eclipse is the result of an attempt of some creature in the sky to swallow the luminary. In order to compel the monster to disgorge it, the people dance round a smouldering fire of old clothing and hair, the stench of which, rising to his nostrils, is expected to cause him to sneeze and disgorge the moon."[11]

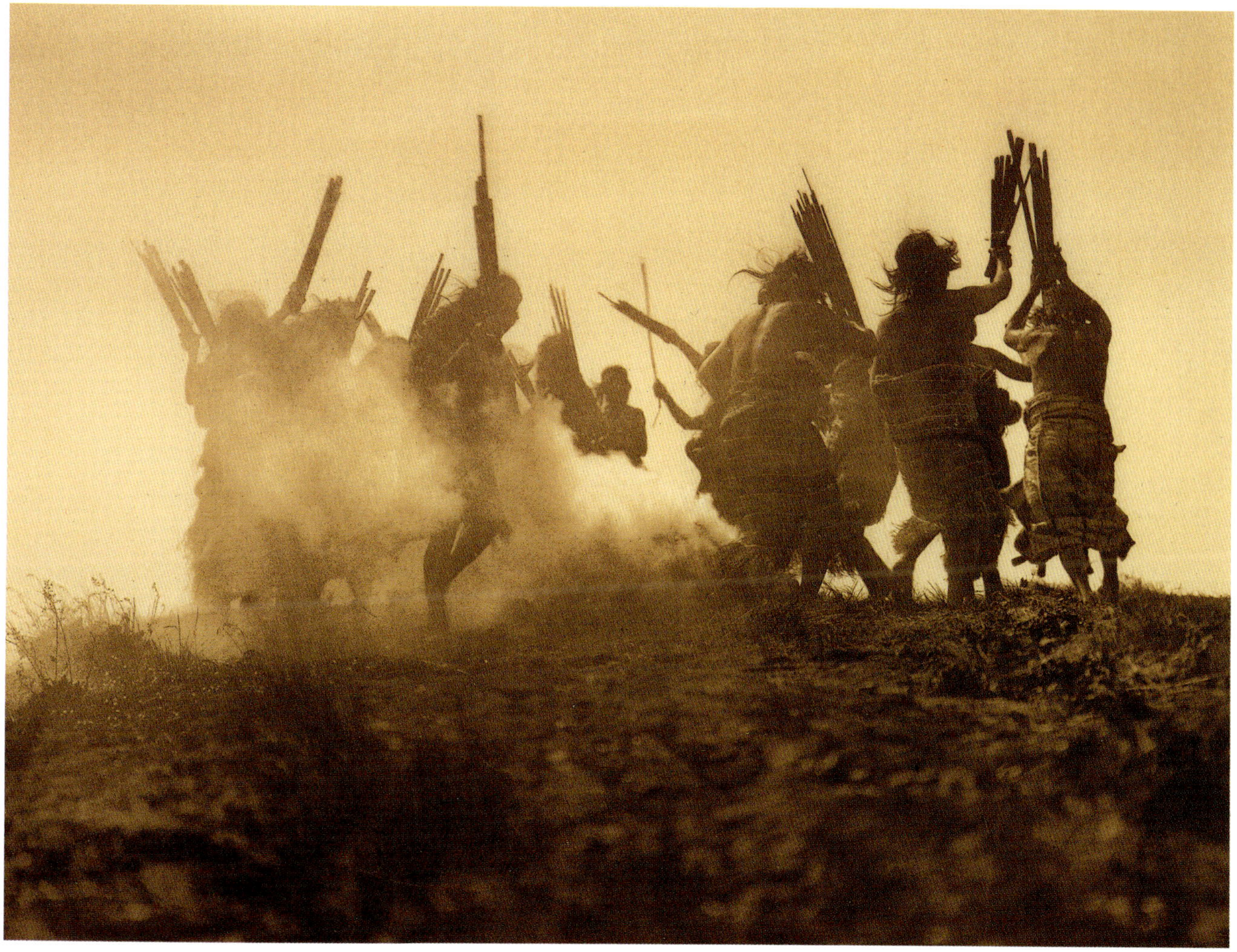

Edward Sheriff Curtis. Plate 355, *Dancing to Restore an Eclipsed Moon-Oágyuhl.* 1915. Photogravure on tissue.
Museum purchase with funds provided by First Interstate Bank of Texas, N.A., at "One Great Night in November, 1994."

Luis González Palma

Luis González Palma, an architect and photographer, brings to his work the skills of construction and model building. Most of his staged scenes blend popular images from his native Guatemala. *La Luna* employs the crescent moon from "loteria" cards, a Latin American game similar to bingo. Binding the woman to the moon refers to the power of the moon's gravity over the Earth's seas and its purported influence on women's menstrual cycles.

Adopting Space Imagery

As a result of his activities with Experiments in Art and Technology (E.A.T.), an organization that actively promoted collaborations between artists and scientists, Robert Rauschenberg was invited by NASA to observe the launching of Apollo 11 in July 1969. This experience inspired a series of thirty-three prints, titled *Stoned Moon,* which Rauschenberg began shortly thereafter at the Gemini, G.E.L. press. The artist incorporated into his compositions many of the photographs and materials that NASA had supplied.

Just as the voyage to the moon required teamwork, the *Stoned Moon* series required intense artistic collaboration and technical innovation on the part of Rauschenberg and the printers at Gemini. From July 1969 to July 1970, the presses were often running fourteen- to sixteen-hour daily shifts.

At the time, *Sky Garden* was the largest single-sheet lithograph ever made—an appropriate size considering the achievement it commemorates. The vibrant red suggests the powerful blast-off of the rocket shown in diagram. The composition includes views of the launch pad, the space center, a bird indigenous to Florida, and images of people to suggest the complexities and intricacies that make a launch possible.

Luis González Palma. *La Luna.* 1991. Gelatin silver photograph on wooden mount. The Museum of Fine Arts, Houston; museum purchase with funds provided by the Photography Accessions Subcommittee in honor of Anne Wilkes Tucker.

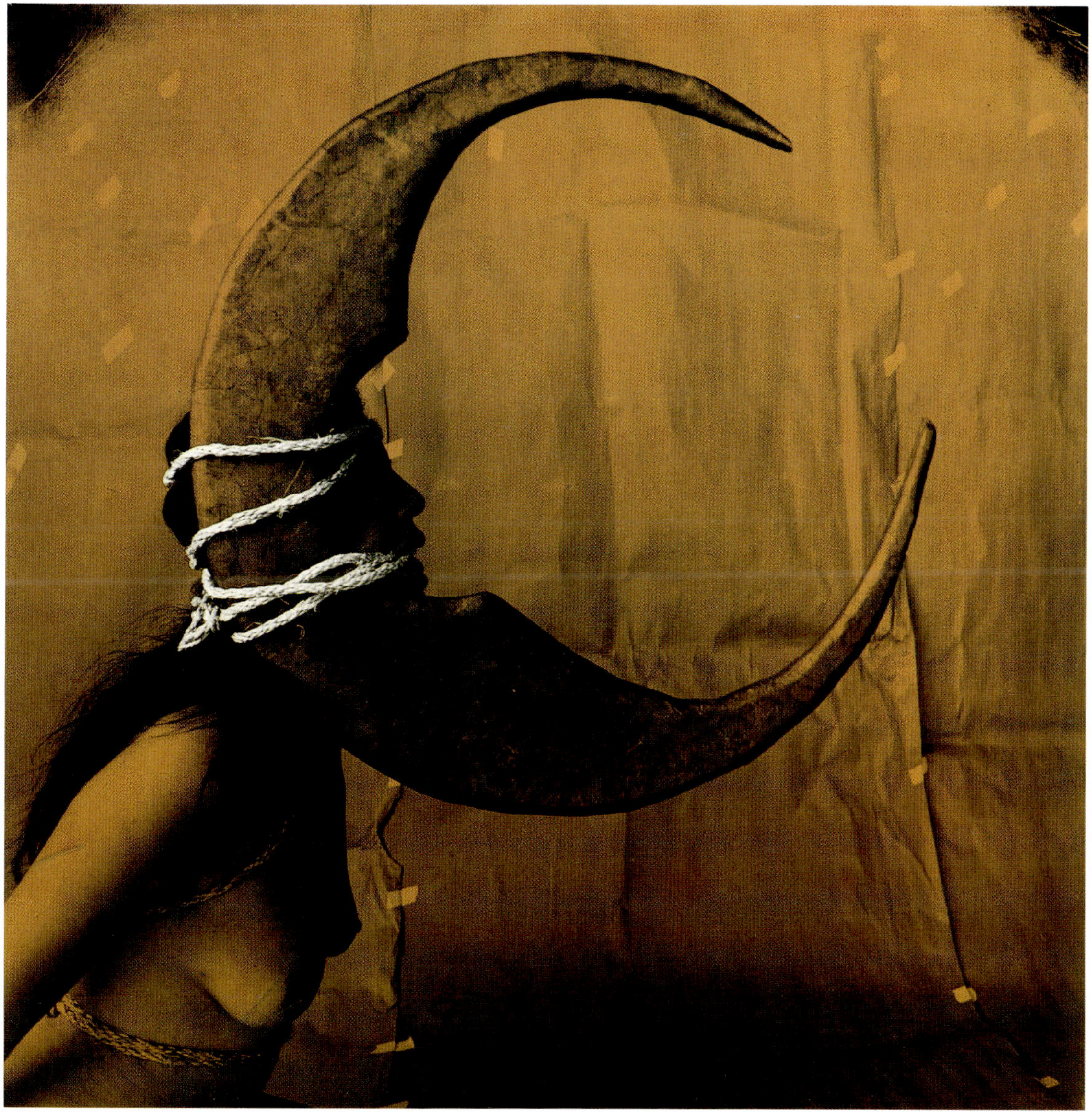

Fact or Fiction

Historian William Ivins, Jr. once wrote, "The nineteenth century began by believing that what was reasonable was true and it would end up by believing that what it saw a photograph of was true."[12] Today, we might say that the twentieth century began by believing that what it saw in a photograph was true and concluded by questioning the truth of any photographic document.

The capacity of computers to seamlessly alter a photograph has completely changed our capacity to trust the photograph as a document. Actually, convincing photographic alterations have always been possible, but computers have made the task faster and have made alteration both undetectable and increasingly realistic.

Photographers have staged events to be photographed ever since the medium's invention—even the simplest studio portrait is a staged event. But the trend to stage obviously fictional events in a realistic manner is a relatively new event in art, as is the desire to fake historical events and make conjured photographic documents convincing.

Thierry Urbain has created a series of photographs envisioning the look of the ancient city of Babylon as discovered much later by imaginary archaeologists. As part of that series he photographed the skies over Babylon, but rather than actually aiming his camera skyward, he photographed grains of sand on a black background. He simulated the look of the sky. Joan Fontcuberta "recognized" the look of the cosmos in the debris on his windshield.

It is not important to either of these photographers that what they are presenting is not factually related to the titles they have given their works. They do not intend to make precise documents, but rather to give form to their imaginations. That they are using photography rather than painting to accomplish this adds a twist to their presentation. Although we know, in today's computer age, that photographs can and do "lie," we recognize nonetheless that both photographers are still depending on a residual element of photography's documentary powers.

Joan Fontcuberta. *Beta Cephei* from the *Constellation* series. 1993. Silver dye-bleach photograph. Courtesy of Zabriskie Gallery, New York, and the photographer.

NOTES

1. M. Susan Barger and William B. White, *The Daguerreotype: Nineteenth-Century Technology and Modern Science* (Washington, D.C.: Smithsonian Institution Press, 1991), 84.

2. Jacques-Henri Lartigue, *Diary of a Century* (New York: The Viking Press, 1970).

3. Quoted in Alan Shepard and Deke Slayton with Jay Barbree and Howard Benedict, *Moon Shot: The Inside Story of America's Race to the Moon* (Atlanta: Turner Publications, 1994), 38-39.

4. Ibid, 133.

5. Ibid, 172.

6. Ibid, 311.

7. Frank White, *The Overview Effect: Space Exploration and Human Evolution* (Boston: Houghton Mifflin Company, 1987), 234.

8. "Remembering Apollo," *Discover* 15, 7 (July 1994), 40.

9. Shepard, *Moon Shot*, 310.

10. White, *The Overview Effect*, 240.

11. *The North American Indian* (New York: Johnson Reprint Corporation, 1970), vol. 10, pl. 355.

12. William M. Ivins, Jr., *Prints and Visual Communications* (Cambridge, Massachusetts: Harvard University Press, 1953), 180.

Selected Bibliography

Bedini, Silvio A., Wernher von Braun, and Fred Whipple. *Moon: Man's Greatest Adventure.* Edited by Davis Thomas. New York: Harry N. Abrams, Inc., 1970.

Bond, Peter. *Heroes in Space: From Gagarin to Challenger.* New York and Oxford, England: Basil Blackwell, 1987.

Chaikin, Andrew. *A Man on the Moon: The Voyages of the Apollo Astronauts.* New York: Viking Press, 1994.

Shepard, Alan, Deke Slayton with Jay Barbree and Howard Benedict. *Moon Shot: The Inside Story of America's Race to the Moon.* Atlanta: Turner Publications, 1994.

White, Frank. *The Overview Effect: Space Exploration and Human Evolution.* Boston: Houghton Mifflin Company, 1987.

Wolff, Tom. *The Right Stuff.* New York: Farrar, Straus & Giroux, 1979.